Fangming Jiang
Jiliang Chen
Wenjiong Cao

Numerical Modeling and Well Layout Design for EGS

Fangming Jiang
Jiliang Chen
Wenjiong Cao

Numerical Modeling and Well Layout Design for EGS

LAP LAMBERT Academic Publishing

Impressum / Imprint
Bibliografische Information der Deutschen Nationalbibliothek: Die Deutsche Nationalbibliothek verzeichnet diese Publikation in der Deutschen Nationalbibliografie; detaillierte bibliografische Daten sind im Internet über http://dnb.d-nb.de abrufbar.
Alle in diesem Buch genannten Marken und Produktnamen unterliegen warenzeichen-, marken- oder patentrechtlichem Schutz bzw. sind Warenzeichen oder eingetragene Warenzeichen der jeweiligen Inhaber. Die Wiedergabe von Marken, Produktnamen, Gebrauchsnamen, Handelsnamen, Warenbezeichnungen u.s.w. in diesem Werk berechtigt auch ohne besondere Kennzeichnung nicht zu der Annahme, dass solche Namen im Sinne der Warenzeichen- und Markenschutzgesetzgebung als frei zu betrachten wären und daher von jedermann benutzt werden dürften.

Bibliographic information published by the Deutsche Nationalbibliothek: The Deutsche Nationalbibliothek lists this publication in the Deutsche Nationalbibliografie; detailed bibliographic data are available in the Internet at http://dnb.d-nb.de.
Any brand names and product names mentioned in this book are subject to trademark, brand or patent protection and are trademarks or registered trademarks of their respective holders. The use of brand names, product names, common names, trade names, product descriptions etc. even without a particular marking in this work is in no way to be construed to mean that such names may be regarded as unrestricted in respect of trademark and brand protection legislation and could thus be used by anyone.

Coverbild / Cover image: www.ingimage.com

Verlag / Publisher:
LAP LAMBERT Academic Publishing
ist ein Imprint der / is a trademark of
OmniScriptum GmbH & Co. KG
Heinrich-Böcking-Str. 6-8, 66121 Saarbrücken, Deutschland / Germany
Email: info@lap-publishing.com

Herstellung: siehe letzte Seite /
Printed at: see last page
ISBN: 978-3-659-64290-6

Numerical Modeling and Well Layout Design for EGS

Fangming Jiang[*], Jiliang Chen, Wenjiong Cao

Laboratory of Advanced Energy System, CAS Key Laboratory of Renewable Energy, Guangzhou Institute of Energy Conversion, Chinese Academy of Sciences (CAS)，Guangzhou 510640, China

Abstract

We first present with great details a three-dimensional transient model of enhanced geothermal systems (EGS) heat extraction processes. This model takes the reservoir as an equivalent porous medium while considers the local thermal non-equilibrium between solid rock matrix and fluid flowing in the fractured rock mass. One other salient feature of this model is its capability of simulating the complete subsurface thermo-hydraulic process in EGS, not only the thermo-flow in the reservoir and well boreholes, but also the heat transport in rocks enclosing the reservoir. Simulation results unravel the underlying mechanism for preferential flow or short-circuit flow forming in well-fractured, homogeneous reservoirs of different permeability values. EGS performance is found to be tightly related to the flow pattern in the reservoir. Thermal compensation from rocks surrounding the heat reservoir contributes little to heat the heat transmission fluid if the operation time of an EGS is shorter than ten years. We then perform a thorough numerical study to the effects of well layout on EGS heat extraction. We find from simulation results that simply deploying more production wells does not necessarily improve the EGS heat

1

extraction performance; an EGS with triplet well layout can perform better than an EGS with a quintuplet well layout or worse than an EGS with the standard doublet well layout. One more finding is an EGS with the injection well positioned close to the edge of the reservoir gets more thermal compensation from the un-fractured rocks surrounding the reservoir during heat extraction. Accordingly, we deduce an optimized EGS well layout must ensure enough long major flow path and less preferential flow in the reservoir, and the injection well is located close to the edge of the reservoir. Last, we discuss about the hot dry rock (HDR) heat recovery factor based on numerous simulated cases and estimate the amount of HDR geothermal resource that can be converted into electricity by EGS.

Keywords: Enhanced or engineered geothermal system; Hot dry rock; Heat extraction; Porous heat reservoir; Local thermal non-equilibrium; Well layout; System optimization design; Numerical model.

Contents

1 Introduction

The earth, from its core to crust, is abundant in renewable heat energy. It is estimated that the total heat stored within the upper 5 km of the earth is about 140×10^6 EJ [1], sufficient for about 2000 years of energy supply to the whole world if considering only 1% of the total heat stored is exploited. Geothermal energy has advantages over solar and wind systems, e.g. steady and continuous energy supply, and is thus one of the few renewable energy resources that can provide base-load power with minimal environmental impacts [2, 3]. The geothermal energy usually talked about is referred to relatively shallow-depth sources, normally in the hydrothermal form. Actually, the shallow-depth hydrothermal energy is only a small amount, less than 10%, of the total accessible geothermal energy. One other form of geothermal energy is located in considerable depths, commonly within subsurface 3-10 km, and in contrast to the hydrothermal energy, there does not exist water, but hot dry rock (HDR). The quantity of HDR heat is enormous. It is reported that the total HDR heat in the United States (US) is about 13×10^6 EJ, a huge amount compared with the annual energy consumption (about 100 EJ) in the US in 2005 [4].

Compared with the hydrothermal heat, the HDR heat is more ubiquitous and commonly of higher temperature, therefore more suitable for electricity generation. Over the last several years there has been much activity to harness the HDR heat. The enhanced or engineered geothermal system (EGS), as an emerging geothermal utilization technology, has attracted broad attention in many countries around the world [5-9]. The concept of EGS, as illustrated in Fig. 1, is relatively new, first proposed by a group of US scientists in the early 1970s. Since then, research projects aimed at developing techniques for the creation of geothermal reservoirs and for the demonstration of power generation have been and are still being conducted around the world. Very recently, the US Department of Energy (DOE) announced the successful grid supply of their first commercial EGS power plant (Desert Peak II) located in Churchill County, Nevada [10].

Unlike the conventional hydrothermal resources usually stored in some shallow-depth permeable sedimentary rocks where quite a large amount of hot fluid is trapped [4, 12], the HDR heat is generally stored in some low-permeability crystalline rocks. To extract a commercial amount of heat from HDR by EGS, fracture stimulations are often required to increase the well productivity [4, 11, 13]. Though stimulation technologies have been successfully deployed in the gas and oil industry for decades and many EGS field tests have been conducted over the past 40 years, the permeability of the created EGS artificial reservoir is not adequate [4, 13]. Creating an artificial reservoir of sufficiently large volume with well-structured fracture network is still a big challenge to the existing fracture stimulation technologies. One major reason is the HDRs are normally much harder and hotter than those in the fossil industry [14]. New technologies of borehole drilling and particularly fracture stimulation are critically needed in the development of commercially viable EGS [4, 14, 15].

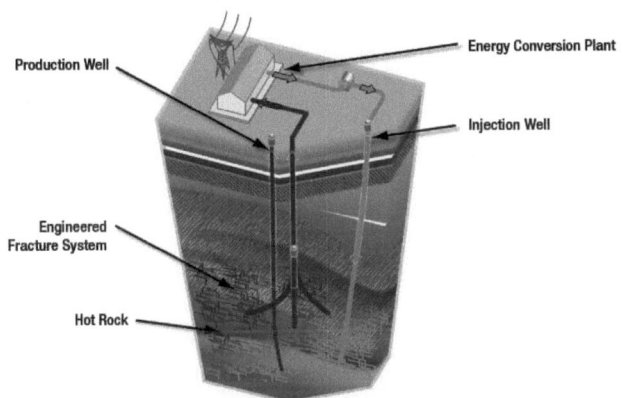

Figure 1 EGS cutaway diagram showing its basic underground and earth-surface components [11].

The morphology and spatial distribution including the connectivity of fractures in the subsurface reservoir significantly affect the performance of EGS. EGS field tests have unraveled that the stimulated fracture network largely depends on the pre-existing fractures and the rock in-situ stress state [4, 16, 17]. A different stimulation treatment may result in fractures of different geometrical dimensions (i.e., aperture, width, and length) [18, 19]. Among the existing stimulation technologies, hydraulic fracturing [18, 20] and chemical stimulation (e.g. matrix acidizing) [18, 21] are the two popular stimulation treatments, whereas thermal stimulation, as one less-noticed stimulation technology, its stimulation effect may be underestimated [22, 23]. It was found that the permeability of some part of the hot rock could be elevated as much as two orders of magnitude when cold water was injected into a high-temperature reservoir [22].

The construction and operation of EGS can induce seismicity [24]. Though lots of works have been done, the mechanisms leading to seismicity remain unclear [24,

25], especially the reasons why some large seismicities (or even the largest) occur after the injection of high pressure fluid [26-28]. It is speculated that the thermal stress caused by the temperature difference between the injected fluid and HDR may be a major mechanism of the induced seismicity, especially for those micro earthquakes occurring after the hydraulic stimulation treatment [24, 25]. If the induced seismicity cannot be avoided or controlled under a threshold magnitude, large scale EGSs can only be sited in rural regions, far away from urban centers, which may deteriorate the economic performance of EGS due to the increased cost of electricity transmission [3, 29]. Besides the thermal stress caused by the rock contraction in response to the injection of cold fluid, the induced or triggered seismicities are related with many other factors, such as the in-situ geological structure, the local rock stress state, the pressure and flow rate of the injected fluid, and chemical reactions between the rock and fluid [24-28].

During EGS operation, the thermo-hydraulic process in the created or enhanced subsurface heat reservoir is crucial to EGS performance, e.g. its capacity (thermal output) and sustainability (lifetime). Keeping the production temperature sufficiently high is important to EGS capacity [4, 11]. During reservoir stimulation, some major fracture zones might form [16, 30, 31], which may cause short-circuit or preferential flow of heat transmission fluid, meaning most of the injected fluid flows into the production well only by several major paths. In this way a large portion of the hot rock in the heat reservoir has little chance to access the circulating fluid, leading to a remarkable dropdown at the production temperature in a short time-duration and thus deteriorating the EGS sustainability. Short-circuit or preferential flow was observed in many of the current operating EGS projects [4, 7, 31-33]. During long-time operation of EGS, interactions between the rock and fluid, such as mineral dissolution from the rock by the fluid or mineral deposition from the fluid on the rock may alter or damage the fracture morphology and/or network, which can also bring significant effects on the heat extraction performance and sustainability of EGS [34-36]

To thoroughly understand the reservoir stimulation, to explain the formation of seismicity, and to unravel the underlying fundamentals of subsurface thermo-hydraulic process, one has to resolve the involved thermal- hydraulic- mechanical- chemical (THMC) coupled processes occurring in the EGS subsurface fractured rock mass and injection/production wells.

Numerous R&D works on EGS have been done since 1970s, which can be mainly classified into three categories: field test [6-8], laboratory experiment [37-39] and numerical modeling [40-45]. Field test is the most direct and convincible way, from which practical engineering experience and geological/geochemical/geophysical data can be obtained. Results got from field tests can be used to validate the other two approaches. However, EGS field tests, including the construction of EGS, are very time-consuming and extremely costly [4]. The obtained practical engineering experience is generally not applicable to one other EGS located at a site of distinct geological/ geochemical/ geophysical conditions. Laboratory experiment is relatively simple and more controllable. Laboratory system acts like a small scale EGS platform and the tests can also provide basic parameters for numerical models. Using experimental results to validate and calibrate numerical models is convenient and, to some extent, very effective. Compared with field tests, laboratory tests are not that costly and can be easily adjusted to perform EGS tests under various geological/ geochemical/ geophysical conditions. During EGS operation, there undergo involved THMC coupled processes in the heat reservoir, physical properties of the rock matrix and circulating fluids, the permeability and porosity in the heat reservoir, and the deep underground geology are all location-dependent and may change with time [41]. Laboratory-scale experiments may be difficult to reflect some real scenarios, especially for the long-term effects of some factors, such as the mineral deposition and/or dissolution. Numerical modeling, as a powerful and highly-efficient tool, has been widely applied in the R&D of EGS [4]. Numerical models, once validated by experiments, can be used to reveal the underlying fundamentals, to evaluate the overall performance of EGS, and to design and optimize the construction and operation of EGS.

The characteristic size of an EGS subsurface reservoir is usually in the order of kilometer, while the fracture aperture is within a few millimeters or even smaller [4, 13]. During EGS operation, heat transmission fluids flow through the narrow fissures, heat exchange occurs at the tortuous rock-fluid interface relying mainly on convective heat transfer, and heat in the solid hot rock matrix is conducted toward the rock-fluid interface. The heat extraction process in EGS is of evident multi-scale and multi-disciplinary nature. Reservoir models are roughly divided into two classes [15,

16]: discrete fracture network (DFN) models and continuum models. DFN models are commonly based on mesoscopic reconstruction of the heat reservoir fracture network, it can account for the effects of an individual fracture on the fluid flow and heat transport. However, the fractures in the reservoir are normally of very complicated configuration/network. Vast data are needed to accurately reconstruct a discrete fracture network of reservoir scale (~km^3). Obviously, direct borehole measurements, due to technical difficulties and costs, are not able to provide this huge amount of data so far [4, 46]. The application of DFN models are also restricted by their intensive demand of computational resources [47]. Continuum models treat the heat reservoir as an equivalent porous medium and do not strictly distinguish where fracture network or rock-matrix is. That is to say, the heat reservoir is characterized by some macroscopic properties (e.g., porosity and permeability etc.) without considering the detailed information on fracture morphology and location. Continuum models have been widely applied in the R&D of EGS nowadays due to its concise physical concept and no need of that huge amount of constitutive data as DFN models [48].

In terms of physical description to the heat reservoir, there are mainly three categories of continuum models reported: the single porosity model [40], the dual porosity model [42, 43], and the multi-porosity model [44, 45]. The single porosity model focuses on the overall performance of EGS, and treats the heat reservoir as a homogeneous porous medium and macroscopically averages the geophysical properties (e.g. permeability and porosity) of the rock matrix and the factures.

11

However, the conventional single porosity model implies an instantaneous local thermal equilibrium between the rock matrix and the fluid flowing in the fractures. That is to say, it assumes the rock and the fluid share a same temperature field. In practical EGS operation, a relatively large temperature difference may exist between the rock and heat transmission fluid in some portion or even most of the reservoir due to the injected fluid is much colder than the hot rock [4, 6, 7] and the local thermal equilibrium assumption may thus be violated. The dual porosity model considers the heat reservoir consists of two different porous regions: the low porosity rock mass and the high porosity fracture region, each of which has its own geophysical properties. The convective heat extraction by the heat transmission fluid is approximately represented by the heat exchange between the two porous regions although the local thermal equilibrium assumption is still implemented in each porous region. However, the dual porosity model generally considers some idealized fracture networks with simple fracture morphology, and an accurate specification of geophysical properties to the two regions is difficult as well since no decisive data can be obtained directly from experimental measurements or geological explorations. The multi-porosity model has been proposed to fully mimic the heterogeneity of heat reservoir. It distinguishes the local structure and configuration of fractures in the heat reservoir, and multiple sets (or a distribution) of geophysical properties are specified accordingly.

The flow of heat transmission fluid in the fractured rock mass and the heat transfer occurring therein play a pivotal role to the EGS heat mining process, significantly affecting the thermal output and lifetime of EGS [4]. Some numerical research works have paid much attention on EGS subsurface thermal- and hydraulic- (TH) coupled processes [48]. Focusing on the complete subsurface heat exchange process, this work presents a 3D transient model for simulating long-term heat mining processes of EGS [49, 50]. This model follows the single-porosity model framework and takes the heat reservoir as an equivalent porous medium of a single porosity. However, it adopts the thought of local thermal non-equilibrium between the solid rock matrix and fluid flowing in the fractures and employs two energy equations to describe heat transfer in the rock matrix and in the fractures, respectively, which makes up the major disadvantage (i.e., incapability of modeling local heat exchange between the rock matrix and fluid) of conventional thermal models for porous media. A few relatively earlier works [43, 44] on modeling heat transfer in porous geothermal reservoir also took the local thermal non-equilibrium assumption. Based on simulation results, we attempt to reveal the formation mechanisms of preferential flow probably occurring in the subsurface heat reservoir and to pinpoint some key factors that affect EGS heat extraction performance. In particular, we carry out well layout design for EGS-s aiming to better exploit HDR heat energy.

2 Numerical Modeling

2.1 Physical model

Physically, the EGS subsurface consists of multiple domains: injection and production wells, heat reservoir, and rocks enclosing the heat reservoir (e.g. the base

13

and cap rock). (see Fig. 1) During numerical modeling, we treat the EGS subsurface geometry of interest as a single-domain consisting of multiple sub-regions: i) region 1 represents the heat reservoir; ii) region 2 the rocks enclosing the heat reservoir; iii) region 3 the injection and production wells, as schematically displayed by Fig. 2. Different regions have distinct geo-physical properties. The heat reservoir is looked as an equivalent porous medium, which is assumed to be isotropic and homogeneous in the present work, and is characterized by a single porosity ε and a finite permeability K. The rocks enclosing the heat reservoir are impermeable to fluids, i.e. $\varepsilon =0$ and $K =0$. Heat transfer in this sub-region relies only on heat conduction. The injection and production wells are looked as open channels, i.e. $\varepsilon =1$ and $K =\infty$. This single-domain treatment circumvents typical difficulties about matching boundary conditions between domains in traditional multi-domain approaches and facilitates numerical implementation and simulation of the complete subsurface thermo-hydraulic process in EGS.

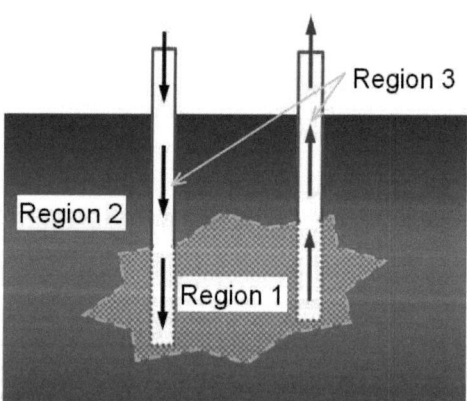

Figure 2   Schematic of the physical model of EGS subsurface geometry. [49, 50]

It is worth pointing out that the assumption of single-porosity porous medium to the heat reservoir can be relaxed for more practical applications, in which the heat reservoir may be heterogeneous and anisotropic. To account for fluid loss during EGS operation, the sub-region 2 can be treated as fluid-permeable porous medium too.

2.2 Mathematical model

The model is focused on modeling and analyses of the subsurface thermo-hydraulic process in EGS. Owing to the single-domain treatment to the EGS subsurface geometry, this model can handle the involved heat transfer and fluid flow with ease. Major assumptions made in the derivation of model equations are summarized as follows.

i) The heat transmission fluid (i.e. water) is in liquid state, i.e. only single phase flow is considered. Due to the large hydraulic pressure (>300 atm), water flowing in the heat reservoir is very possibly in liquid state.

ii) The porous heat reservoir is fluid saturated. That is to say, no immiscible gas is trapped in the rock fractures.

iii) The fluid flow in heat reservoir is in laminar regime and the Reynolds number is sufficiently low to suffice the applicability of Darcy law.

iv) There do not exist any fluid-structure interactions, including chemical dissolution/deposition and physical pressing etc.

The governing equations consisting of a series of conservation equations yield.

Mass continuity equation:

$$\frac{\partial(\varepsilon\rho)}{\partial t} + \nabla\cdot(\rho\mathbf{u}) = 0$$

(1)

Momentum conservation equation:

$$\frac{\partial(\rho\mathbf{u})}{\partial t} + \nabla\left(\frac{\rho\mathbf{u}}{\varepsilon}\cdot\mathbf{u}\right) = -\nabla(\varepsilon P) + \nabla\cdot\mu\nabla\mathbf{u} - \varepsilon\frac{\mu}{K}\mathbf{u} + \varepsilon\rho\mathbf{g}$$

(2)

Energy conservation equation for the heat transmission fluid flowing in the fractures:

$$\frac{\partial\left[\varepsilon(\rho c_p)_f T_f\right]}{\partial t} + \nabla\cdot\left[\varepsilon(\rho c_p)_f T_f\mathbf{u}\right] = \nabla\cdot\left(k_f^{eff}\nabla T_f\right) + ha(T_s - T_f)$$

(3)

Energy conservation equation for heat transport in the rock matrix of the heat reservoir or in the surrounding impermeable rocks:

$$\frac{\partial\left[(1-\varepsilon)(\rho c_p)_s T_s\right]}{\partial t} = \nabla\cdot\left(k_s^{eff}\nabla T_s\right) - ha(T_s - T_f)$$

(4)

The application of the full-form Navier-Stokes equation, Eq. (2), enables a general treatment to the fluid flow in open-channel injection and production wells and in the porous heat reservoir. The third term in the right hand side of Eq. (2) represents the Darcy resistance. We consider local thermal non-equilibrium between the rock matrix and fluid flowing in the fractures of the porous heat reservoir, and thus employ two energy equations, Eqs. (2) and (3) to describe the heat conduction in HDR (or rock matrix) and the heat convection and advection in fluid, respectively. In the energy conservation equations there is a term $\pm ha(T_s-T_f)$ introduced to model the heat exchange between solid rock matrix and fluid flowing in the fractures in the heat reservoir. Bruggeman correction with a correction factor of 1.5 is applied to determine the effective heat conductivity k^{eff}, i.e. $k_s^{eff}=k_s(1-\varepsilon)^{1.5}$ and $k_f^{eff}=k_f\varepsilon^{1.5}$.

2.3 Heat exchange in reservoir

In terms of the present modeling framework, rock-fluid heat exchange in the heat reservoir of EGS is very important to the heat extraction. The heat exchange rate is calculated by the product of specific surface area (a) of fractures, the rock-fluid convective heat transfer coefficient (h), and the rock-fluid temperature difference (T_s-T_f), as expressed in Eq. (3) or (4). The specific surface area of fractures is determined from the geometrical configuration of fractures; the rock-fluid convective heat transfer coefficient depends on the thermo-physical properties of the fluid and the fluid flowing velocity [51]. The product of h and a directly affects the rock-fluid heat exchange rate in the heat reservoir, and dominates the temporal variation of fluid temperature at the outlet of production well. We can roughly estimate h and a based on the so-called parallel plate model, which geometrically approximates the complicated fracture network as N equidistant parallel fractures of aperture d. (Fig. 3) Because normally the fracture length $L >> d$, the characteristic size (hydraulic diameter) d_h of fractures is around $2d$. Simple derivation leads to the following Eqs. (5) and (6).

$$h = \frac{k_f Nu}{d_h} = \frac{k_f Nu}{2d}$$

(5)

$$a = \frac{S_p}{V_{total}} = \frac{2\varepsilon}{d}$$

(6)

Actual or more accurate values of h and a for practical EGSs can be measured experimentally or calculated from fine-scale models.

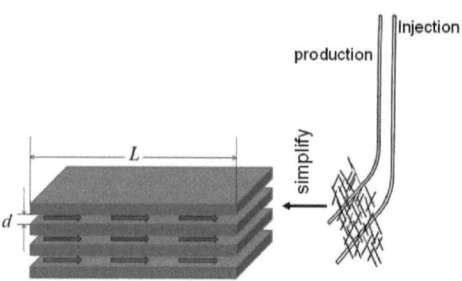

Figure 3   Schematic about the concept of parallel fracture simplification to the heat reservoir

2.4 Numerical strategy

Equations (1), (2), (3), and (4) together with appropriate boundary conditions and initial conditions are solved in the commercial CFD flow solver, Fluent®, which is based on the finite volume approximation. By customizing its flexible User Defined Functions (UDF), various source terms, geo-physical properties, and non-standard advection-convective terms in the governing equations are implemented. The well-known SIMPLE (Semi-Implicit Method for Pressure Linked Equation) algorithm is used to address the pressure-velocity coupling. The first order upwind differencing scheme is generally used for discretization of the spatial-derivative terms and a fully implicit scheme for discretization of the transient terms. To accelerate convergence, the AMG (algebraic multi-grid) iterative method is applied to solve the linearized algebraic equations.

3 Performance Prediction of EGS Heat Extraction

3.1 Case setups

The volume of the created reservoir determines the total thermal energy that is directly exposed to the heat transmission fluid. It is an important factor affecting the commercial viability of EGS. Though no concrete proofs can explicitly show the geometrical configuration of the reservoir, micro-seismicity measurements are able to reveal its approximate volume [17, 33, 52]. Rock-fracturing technologies have already demonstrated its capability of creating up to 3 km^3 reservoirs [4]. The shape of artificial reservoirs is highly irregular and very hard to describe; the boundary between the reservoir and surrounding un-fractured rocks is hard to define because of the spatially gradual change of fracture network caused by, for instance, the hydrothermal alteration effects on the rock permeability and porosity [17, 33, 35, 52]. An imaginary EGS, geometry (including geometrical dimensions) and mesh system being displayed in Fig. 4, is considered. The heat reservoir is a 500×500×500 m volume of rhombus-shape in the *x-y* plane, located approximately at subsurface 4000 m depth. The simulated domain is a 2000×6000×2000 m cuboid. The injection and production wells are both 0.2×0.2 m square-shaped in the *x-z* plane. Structural hexahedral meshes are used to discretize the calculation domain. Owing to the symmetry about the mid-*z* plane, only lower-*z* half of the geometry is simulated. The mesh was elaborately designed to get sufficiently fine resolution in the injection and production wells and in the heat reservoir. The mesh system has around 300,000 numerical elements in total. Grid-independence tests have been conducted to guarantee the present mesh system gives solutions of satisfying accuracy.

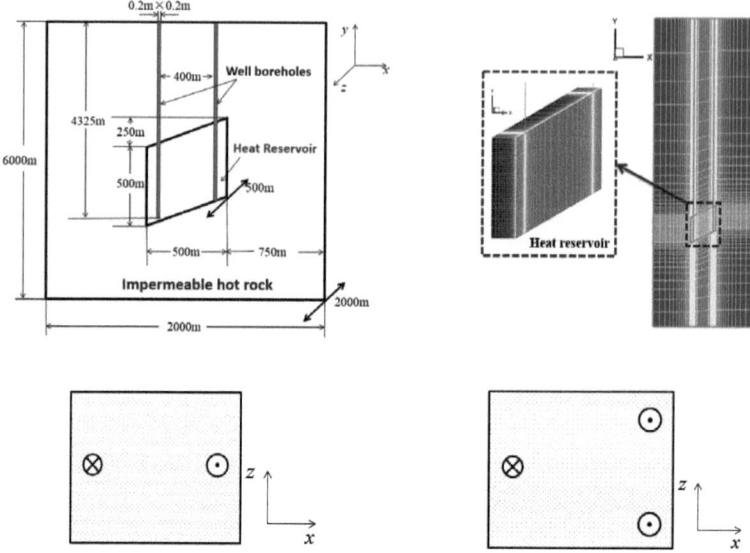

Figure 4 EGS subsurface geometry, mesh system and schematics of doublet and triplet well layouts. y denotes the vertical axis along the depth direction.

Local geothermal gradient is an important factor considered during EGS site-picking. The geothermal gradient of the selected EGS site is usually higher than the average value to avoid too deep well-drilling. In this section, temperature values at the top and bottom x-z plane are fixed at 300 K and 540 K, respectively; the geothermal gradient is assumed to be constant, 4 K/100 m.

Though the suggestion of using supercritical CO_2 as EGS heat transmission fluid has been proposed for years [53], liquid water is still the most common heat transmission medium used in the to-be-developed or in-operation EGS plants. Liquid water is taken as the heat transmission fluid in this section. Thermo-physical properties of fluid and rock are assumed temperature-independent, as listed in Table 1. Initially, the injection and production wells are full of water of temperature 300 K; the temperature of water in the fractures of the heat reservoir is equal to the local rock matrix temperature. The temperature of the injected water is fixed at 300 K all through EGS operation.

Table 1 Thermophysical Properties.

	Thermal capacity (J/kg/K)	Thermal conductivity (W/m/K)	Density (kg/m^3)	Viscosity (kg/m/s)
Fluid	4200	0.6	1000	0.001
Rock	1000	2.1	2650	N/A

Reservoir permeability and porosity may be the two most important parameters dominating EGS heat extraction process. They dictate the flow distribution and the flow resistance (i.e. the needed external pump work) and thereby directly affect the EGS performance, including the heat extraction performance, lifetime, and economic performance etc. Many factors, such as the rock local stress state, in-situ natural fractures, fluid injection pressure, and rock/fluid chemical composition, may play important roles in the engineering of EGS reservoir [21, 54, 55]. The rock temperature evolution during the operation of EGS can also greatly change the reservoir permeability and/or porosity as the hot rock contracts in response to the injected cold fluid [22, 25, 56]. Limited by the current technologies, it is almost impossible to obtain the detailed local distributive information of the permeability and porosity in the reservoir. Literature works suggested overall reservoir permeability and porosity values, while both are very divergent data (even for the same EGS test/demonstration field) ranging from 10^{-4} [40, 57] to 10^{-2} [41, 58, 59] for the reservoir porosity and 10^{-8} [60] to 10^{-18} m^2 [61, 62] for the reservoir permeability. The heat reservoir is assumed to be homogeneously fractured with a constant porosity of 0.01 for all the cases considered, while the permeability in the reservoir is varying within 10^{-14}–10^{-8} m^2 in this section. All the cases we considered for conducting a systematic parametric study to EGS heat mining process are tabulated in Table 2.

It is reported that a commercially acceptable flow rate of heat transmission fluid is about 80 kg/s [4]. We consider a fluid flow rate of 150 kg/s for all the cases in this section. The product of h and a is actually the constitutive condition for heat exchange between the fluid flowing in the fractures and the rock matrix in the reservoir. The ha is taken to be 1.0, 2.0, or 5.0 $W/m^3/K$ for the cases considered in this section. Additionally, we set up a special case, case 7, which has infinite ha value and is actually calculated with the conventional thermal equilibrium model. The heat conductivity of the solid rock is taken to be 2.1 W/m/K. If we artificially specify the heat conductivity of the impermeable rock surrounding the reservoir as zero (case 8), the thermal compensation from the surrounding rocks is excluded. To study the well layout effects on EGS heat extraction process, we consider two more cases, cases 9 and 10, which have the triplet well layout as illustrated in Fig. 4.

Table 2 Cases studied in Section 3.

Case #	Reservoir permeability (m^2)	ha $(W/m^3/K)$	Thermal compensation	Well layout
Case 0	10^{-12}	1.0	Considered	doublet
Case 1	10^{-8}	1.0	Considered	doublet
Case 2	10^{-9}	1.0	Considered	doublet
Case 3	10^{-10}	1.0	Considered	doublet
Case 4	10^{-14}	1.0	Considered	doublet
Case 5	10^{-12}	2.0	Considered	doublet
Case 6	10^{-12}	5.0	Considered	doublet
Case 7	10^{-12}	∞	Considered	doublet
Case 8	10^{-12}	1.0	Not considered	doublet
Case 9	10^{-8}	1.0	Considered	triplet
Case 10	10^{-12}	1.0	Considered	triplet

In the following sub-sections 3.2-3.6, we will present simulation results to analyze and unravel the formation mechanism of preferential flow in EGS subsurface porous heat reservoir, and to discuss and study effects of the reservoir permeability, the constitutive heat exchange condition (i.e. ha) in the reservoir, the thermal compensation from rocks surrounding the reservoir, and the well layout on EGS performance. To facilitate analyses and discussion, we define three parameters for the evaluation of EGS performance as follows.

23

1) Production temperature $T_{f,out}(t)$: the fluid temperature at the outlet of production well.

2) EGS service-time or lifetime τ: the time-duration for an EGS operating until the production temperature $T_{f,out}(t)$ declines down to 423.15 K, i.e. 150 ℃.

3) Local heat extraction ratio $\gamma_L(t)$: the extracted heat divided by the stored or maximum extractable heat. Since the heat capacity and density of rock are assumed constant, the definition of γ_L is expressed as

$$\gamma_L(t) = \frac{T_s(0) - T_s(t)}{T_s(0) - T_{f,in}}$$

(7)

where $T_s(0)$, $T_s(t)$ and $T_{f,in}$ represent the initial rock temperature, the rock temperature at time instant t, and the fluid injection temperature, respectively.

3.2 Preferential flow in reservoir

Preferential flow is usually detrimental to EGS performance. With the occurrence of preferential flow, a large quantity of heat transmission fluid flows along a few preferential paths, leading to a quick dropdown at the production temperature while the heat stored in a large portion of the reservoir has not yet been extracted out.

Aiming to unravel the formation mechanism of preferential flow in EGS subsurface porous heat reservoir, we analyze the fluid flows for cases with different permeability in the reservoir. Figure 5 presents the calculated x-velocity distribution in the heat reservoir for four cases. These cases, cases 1, 3, 0 and 4 have permeability in the reservoir of 1.0×10^{-8} m^2, 1.0×10^{-10} m^2, 1.0×10^{-12} m^2, and 1.0×10^{-14} m^2,

24

respectively. Three contour plots with the monitoring planes set at the reservoir symmetric *x-y* plane, a diagonal plane of the reservoir, and an *x-z* plane located approximately at the mid-depth of the reservoir, respectively, are displayed for each case.

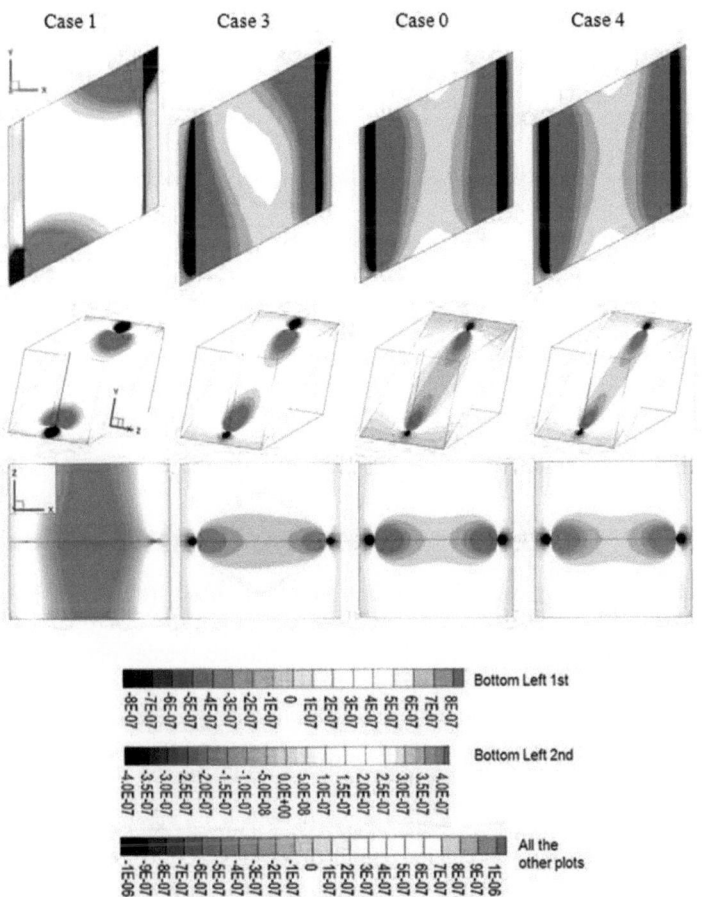

Figure 5 *x*-velocity (m/s) distribution on the reservoir symmetric *x-y* plane (top row), a diagonal plane (mid row) of the reservoir, and an *x-z* plane located approximately at the middle depth of the reservoir (bottom row) for doublet well cases of different reservoir permeability.

Obviously, all the cases show some extent preferential flow. For case 1, the preferential flow is forming along the bottom of the injection well to the top of the production well; in the depth direction (y) the preferential flow is evident whereas in the z direction the fluid flow is pretty uniform. For case 0, an evident preferential flow is seen in the z direction, the fluid prefers to flow within a narrow-z region centering about the reservoir symmetric x-y plane; in the y direction the fluid flow looks more uniform. The scenario for case 3 is in-between of case 0 and 1. Comparing case 4 with case 0, we find that the flow patterns are pretty much the same, indicating the reservoir permeability has little effect on the fluid flow pattern when it is below 1.0×10^{-12} m^2.

During fluid circulation, the gravity effect and the transformation between hydrodynamic and hydrostatic pressure result in such a basic pressure difference in the reservoir with the highest pressure position located at the vicinity of the injection well bottom and the lowest pressure position near the production well exiting from the reservoir. For a case with high reservoir permeability (e.g. case 1) the Darcy resistance is much smaller than the aforementioned basic pressure difference. Therefore, a preferential flow is seen to form along the opposite direction of the basic pressure gradient, from the bottom of injection well to the top of production well in the reservoir, as seen from Fig. 5. The Darcy resistance is inversely proportional to the reservoir permeability. For the cases with lower reservoir permeability (e.g. case 0 and 4), the largely increased Darcy resistance dominates over the effect of the basic pressure difference. For any fluid flow across the reservoir, regardless of where it starts from the injection well or enters into the production well, the total driven pressure difference of fluid flow is approximately the same. Therefore, no apparent preferential flow is seen in the *y* direction any more, while in the *z* direction, distinct flow distance causes severe preferential flow, as evident by Fig. 5.

3.3 Effects of reservoir permeability

The reservoir permeability for case 0 and case 1 have four orders of magnitude difference. Figure 6 presents the rock temperature evolution during EGS operation from these two cases. The rock mass near the injection well is seen to be first cooled down and the low temperature region is gradually extending along the preferential flow path (refer to Fig. 3). The rock temperature evolution closely relates to the fluid flow pattern in the reservoir for both cases. Note that a dark line (line AB) is plotted in Fig. 6 for both cases, which designates the major preferential flow path in the reservoir. Line AB will be used to create the line graphs shown in Fig. 7.

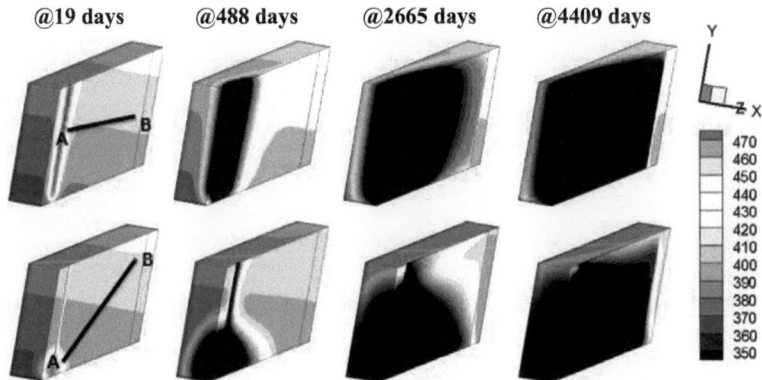

Figure 6   Temperature (K) evolution of rock in heat reservoir. Upper row: case 0; lower row: case 1.

One advantage of the present model over the conventional thermal model for porous heat reservoir is its capability to give the fluid and rock temperature both, thus enabling analyses of local heat exchange in the reservoir. Since the *ha* is assumed constant in the present work, the rock-fluid temperature difference actually indicates the local heat extraction rate. Figure 7 depicts the rock-fluid temperature difference profiles along line AB at a few representative time instants during EGS operation for cases 0 and 1. Both cases have similar profiles of rock-fluid temperature difference along line AB at comparable time instants examined. During the early period of EGS operation (say, 0-19 days), the injected cold fluid is quickly heated up rightly after it flows into the reservoir duo to the large rock-fluid temperature difference; the heat exchange process mainly occurs in the region near the injection well, indicated by almost zero rock-fluid temperature difference at positions of large distance to position A. As the heat extraction progresses, the rock near the injection well cools down and the region where the fluid can have the highest heat exchange rate with the rock shifts toward the production well side, indicated by the peak value of rock fluid temperature difference shifts toward position B. With further progress of the heat extraction process (e.g., after about 992 days for case 0 and 2665 days for case 1), the fluid flows into the production well without being fully heated-up, indicated by a positive non-zero rock-fluid temperature difference at position B. Due mainly to the overall decreasing rock temperature, the peak value of rock-fluid temperature difference is decreasing throughout the EGS operation.

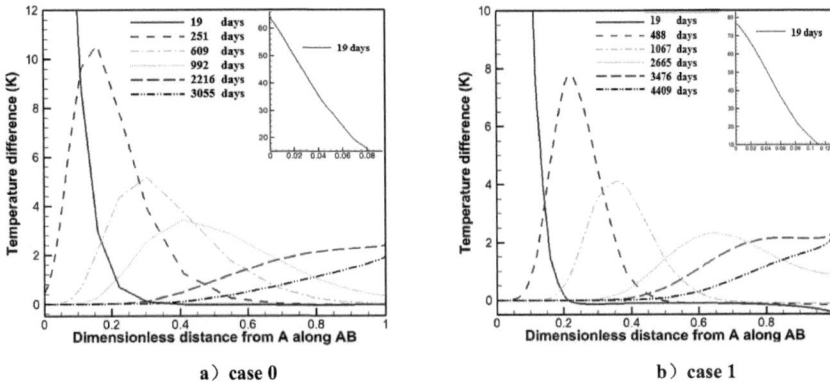

a) case 0   b) case 1

Figure 7   Rock-fluid temperature difference along line AB at different operation time.

It is worth pointing out that the rock-fluid temperature difference can be as large as 10 K seen from Fig. 7, particularly during the very early period of EGS operation its value can be tens of Kelvin degrees, which corroborates the necessity of utilizing the thermal non-equilibrium model.

EGS production temperature and lifetime (or service-time) are two important factors for the evaluation of EGS performance [63, 64]. Figure 8 shows the production temperature evolution curves for a series of cases of different reservoir permeability. For all these cases, the production temperature increases first, and then decreases with time. In the present work, 4 K/100m geothermal gradient is assumed, i.e. the rock temperature at the reservoir bottom is about 20 K higher than that for the rock at the reservoir top edge. The increase at the production temperature is caused by the fluid that has been fully heated-up at the reservoir bottom gradually flows out. For the cases of smaller reservoir permeability (e.g. cases 0, 3, or 4), since the heat extraction in the reservoir depth direction is more uniform, the increase potential of the production temperature is not that large as the other two cases (i.e. cases 1 and 2). The decrease at the production temperature after a certain time into EGS operation is caused mainly by the region of the highest heat exchange rate shifts too close to the production well (see Fig. 7), which leads to the fluid cannot be fully heated-up before entering into the production well.

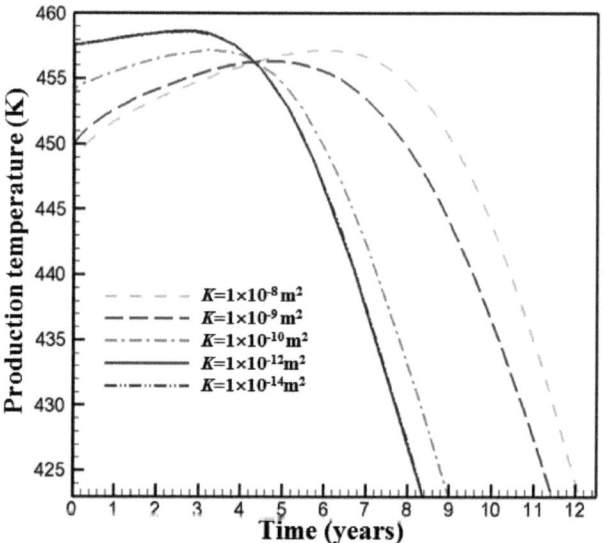

Figure 8   Production temperature evolution curves for doublet well cases of different reservoir permeability.

It is easy to find out from Fig. 8 that the lifetime of EGS decreases with the increase of reservoir permeability. The reservoir permeability generally has very significant effect on EGS lifetime, by increasing the reservoir permeability from 10^{-12} m^2 to 10^{-8} m^2, the EGS lifetime is prolonged from 8.4 years to 12.1 years. For the two cases with reservoir permeability to be 10^{-12} m^2 and 10^{-14} m^2, respectively, there shows seemingly no or little effect of reservoir permeability on the production temperature and lifetime of EGS, agreeing well with the observation from Fig. 5: once the reservoir permeability is lowered below 10^{-12} m^2, further decrease at the reservoir permeability has little effect on the flow pattern in the reservoir. Nevertheless, this does not mean that EGS performance will not benefit from enhancing the reservoir permeability since larger reservoir permeability means smaller flow resistance and less consumption of external pump work [4, 11, 64].

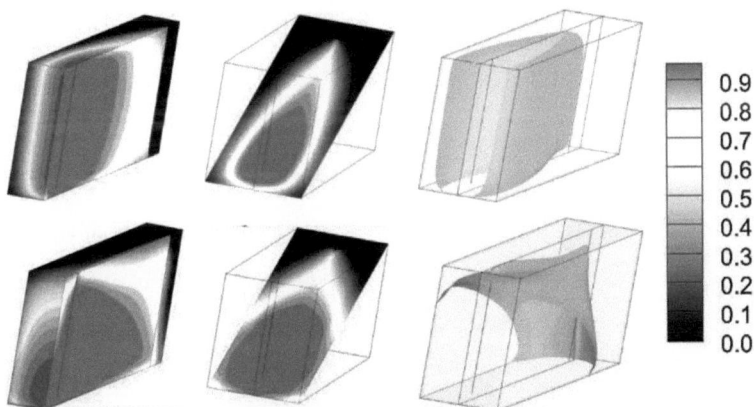

Figure 9 Local heat extraction ratio distribution in heat reservoir at the end of EGS operation for case 0 (upper row) and case 1 (Lower row). Left column: 3D plot; Middle: 2D contour plot on a diagonal plane in reservoir; Right: iso-value surface (γ_L=0.5).

The application of local thermal non-equilibrium model to the heat extraction process of EGS enables scrutinizing the local heat exchange in reservoir. At the end of EGS operation (i.e. the time when the production temperature drops down to 423.15 K), we draw the local heat extraction ratio in reservoir in Fig. 9 for cases 0 and 1. For each case, the 3D distribution, a 2D contour plot on a diagonal plane of reservoir, and the iso-value surface with $\gamma_L = 0.5$ are presented. Altogether, these plots shed light on the local heat exchange in reservoir. We can easily find out from these plots, how the preferential flow affects EGS performance, where the heat extraction has not sufficiently carried out, why the production temperature drops down to the ceasing-operation temperature while a large portion of rock still has high temperature, and etc. All these findings corroborate the observations gotten from the foregoing Figs. 5-8 and are surely helpful to the design and optimization of EGS construction and operation.

3.4 Effects of ha

The specific surface area (a) of fractures and the rock-fluid convective heat transfer coefficient (h) are two important parameters affecting the heat extraction in EGS subsurface heat reservoir as the heat extraction rate is directly proportional to h and a each. The parameter, a, represents the rock-fluid heat exchange area per unit reservoir volume and is purely a geometrical parameter with its value depending on the fracture morphology. Unlike geometrical parameters or material physical properties, the h depends on many more factors including the flow regime, thermal conductivities of rock and fluid, and the surface roughness of fractures, etc. In practice, the values of h and a are both location-dependent and may vary during EGS operation. The accurate determination of h and a value needs to perform experimental measurements or fine-scale numerical modeling. In this work, we take the product of h and a as a single parameter and specify it a constant value (e.g., case 0, $ha = 1.0$ W/m^3/K; case 5, $ha = 2.0$ W/m^3/K; case 6, $ha = 5.0$ W/m^3/K).

Figure 10 depicts the profiles of rock-fluid temperature difference along line AB at various time instants for case 6 (the position of line AB is the same as case 0 shown in Fig. 6). Compared with the curves shown in Fig. 7a, the curves in Fig. 10 generally show diminished rock-fluid temperature difference. The larger *ha* value in case 6 enhances the rock-fluid heat exchange in the heat reservoir and makes the local heat exchange more approach to the thermal equilibrium process. For case 6, due to the enhanced rock-fluid heat exchange, the fluid can be still fully heated-up by the rock after 1060 days into the EGS operation, about 100-day prolongation in comparison with case 0.

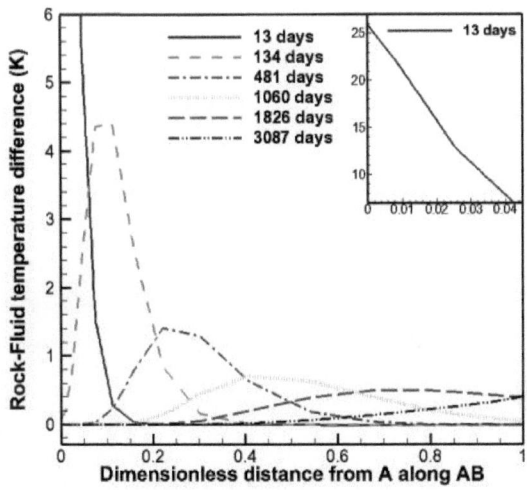

Figure 10 Rock-fluid temperature difference along line AB at different EGS operation time for case 6.

Figure 11 compares the production temperature as a function of EGS operation time calculated from the three cases (i.e., cases 0, 5 and 6) of *ha* value 1.0, 2.0 and 5.0 W/m³/K, respectively. The case of larger *ha* value, shows better performance. Besides the production temperature rises up to a higher maximum value, the EGS lifetime can be prolonged more or less.

Specially, we calculated one case, case 7 ($ha = \infty$), with the conventional thermal equilibrium model. The obtained production temperature as a function of EGS operation time is also provided in Fig. 11. The thermal equilibrium model overestimates the production temperature, but will not result in significant deviations. That is to say, for the predictions of EGS overall performance (e.g. production temperature and lifetime), the thermal equilibrium model is normally adequate. It is worth pointing out that in practical EGS heat reservoirs, which may be not homogeneously fractured as assumed in the present work, h and/or a value can be locally very small, and the application of local thermal equilibrium model may cause relatively large deviations even for the predictions of EGS overall performance.

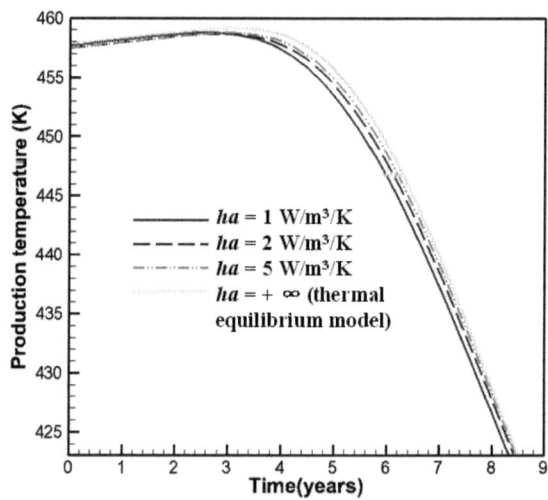

Figure 11   Production temperature as a function of EGS operation time from case 0, 5, 6 and 7.

3.5 Effects of thermal compensation from rocks surrounding the reservoir

The stimulated heat reservoir is a finite volume of fractured rocks, which is normally enclosed by some impermeable solid rocks. During EGS operation, thermal compensation from hot rocks enclosing the heat reservoir may contribute to the heating of heat transmission fluid [65, 66]. The model used in the present work simulates the complete subsurface heat transfer process, automatically handling thermal compensation from rocks surrounding the reservoir. This thermal compensation relies solely on the heat conduction in the surrounding rocks. In case 8, we artificially specify the heat conductivity of surrounding rocks to be zero. In this way, the thermal compensation is completely excluded. The calculated production temperature as a function of EGS operation time from two cases: case 0 (with normal heat conductivity, 2.1 W/m/K, in the surrounding rocks) and case 8, in Fig. 12. The two curves vary almost in the same trace with EGS operation time, indicating the thermal compensation from surrounding rocks has little contribution to the heating of heat transmission fluid. A close inspection of the two highlighted subsets of Fig. 12 finds the thermal compensation even leads to slightly lower production temperature during the early period of EGS operation although for sufficiently long-time EGS operation it does slightly increase the production temperature. Note that for case 0 the examined EGS operation time is around 8.3 years. We further deduce that the contribution of thermal compensation may augment with the EGS operation time, but for an EGS with the operation time-duration shorter than 10 years the thermal compensation may only have limited contribution to the heat extraction process.

During the early period of EGS operation, the negative effect of thermal compensation from rocks surrounding the reservoir on the heat extraction process seems hard to explain. To understand this, with respect to case 0 we draw a contour plot of rock temperature on the symmetric x-y plane at a representative time instant (i.e. 992 days) into the EGS operation in Fig. 13. It is seen that at the vicinity of the production well, the temperature of rock matrix in the reservoir is higher than the temperature of the rocks adjoining the reservoir, meaning some heat is transported from the reservoir out into the surrounding rocks. Although we see some heat is transported from the surrounding rocks into the reservoir at the vicinity of the injection well, the net thermal compensation effect of all the surrounding rocks is negative, leading to a lowered production temperature. Then another question pops up.

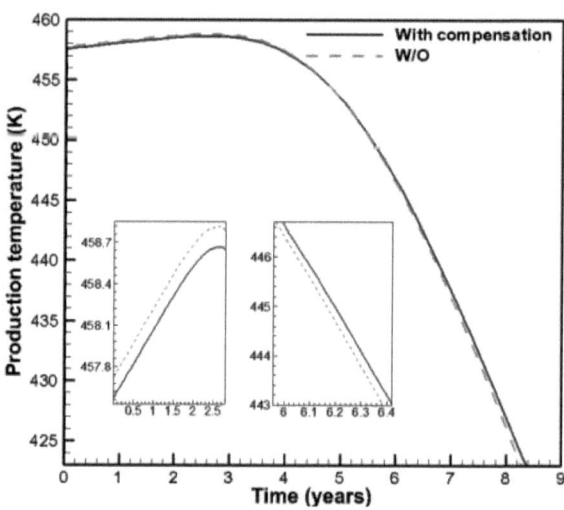

Figure 12   Comparison of production temperature of case 0 and 8.

What causes the increase at the temperature of rock matrix at the vicinity of the production well? As aforementioned, there exists a 4 K/100m geothermal gradient, meaning about 20 K temperature difference along the depth direction in the reservoir, initially. During the early period of EGS operation, for the fluid flowing from the very bottom of the reservoir, since it has been fully heated-up by the hotter rock matrix therein, its temperature is higher than the rock matrix at the vicinity of the production well, leading to an inverse heat extraction from the fluid to rock matrix. This could explain the early period temperature increase for rock matrix at the vicinity of the production well. By carefully checking the original data used to create Figs. 6 and 7, the rock-fluid temperature difference may be slightly negative at positions close to the production well during early period of operation, which corroborates the occurrence of inverse heat extraction therein.

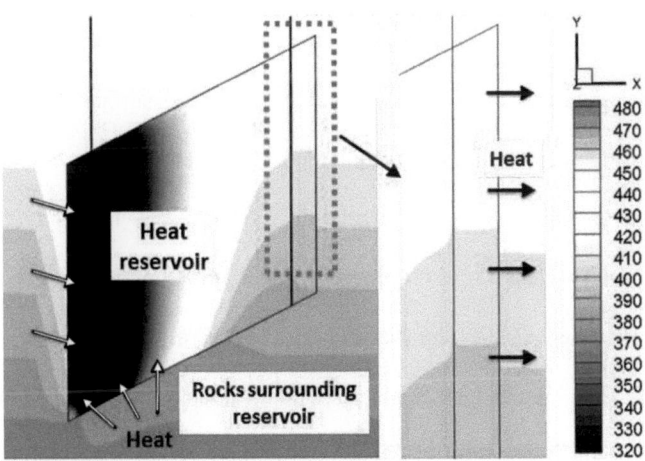

Figure 13 Contour plot of rock temperature (K) on the symmetric *x-y* plane at 992 days, illustrating the early period negative effect of thermal compensation on the heat extraction. Results are from case 0.

3.6 Effects of well layout

Figure 14 presents the calculated x-velocity distribution in the heat reservoir for the two EGS cases: cases 9 and 10, both of triplet well layout (refer to Fig. 4). Cases 9 and 10 have permeability in the reservoir of 1.0×10^{-8} m^2 and 1.0×10^{-12} m^2, respectively. Three contour plots with the monitoring planes set at the reservoir symmetric x-y plane, a diagonal plane of the reservoir, and an x-z plane located approximately at the mid-depth of the reservoir, respectively, are displayed for each case. Compared with the calculated x-velocity distribution (see in Fig. 5) of cases 1 and 0, the triplet well layout largely homogenizes the fluid flow in the reservoir, particularly it effectively restrains the preferential flow in the z-direction of the reservoir.

We draw the local heat extraction ratio in the reservoir in Fig. 15 for cases 9 and 10 at the end of EGS operation. For each case, the 3D distribution, a 2D contour plot on a diagonal plane of the reservoir, and the iso-value surface with $\chi = 0.5$ are presented. We also compare the production temperature as a function of EGS operation time for EGS cases of different well layouts in Fig. 16. As evident by these figures, the triplet well EGSs show improved heat extraction performance compared with the doublet well EGSs. The triplet well EGS of reservoir permeability 10^{-8} m^2 has a lifetime 15% longer than the corresponding doublet well EGS, while the EGS of smaller reservoir permeability 10^{-12} m^2 shows more pronounced well layout effect, the EGS lifetime is prolonged by 48% if changing the well layout from doublet to triplet.

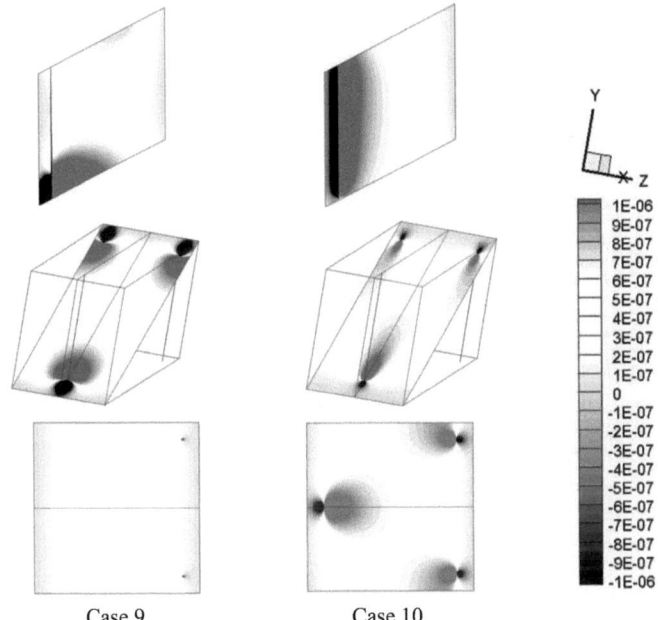

Case 9   Case 10

Figure 14 *x*-velocity (m/s) distribution on the reservoir symmetric *x-y* plane (top row), a diagonal plane (mid row) of the reservoir, and an *x-z* plane located approximately at the middle depth of the reservoir (bottom row) for triplet well cases of different reservoir permeability.

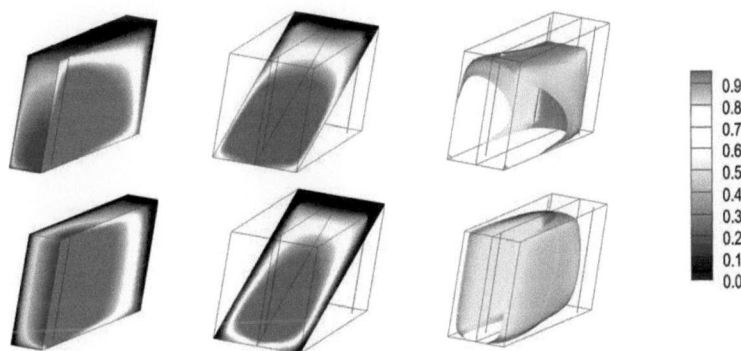

Figure 15 Local heat extraction ratio distribution in heat reservoir at the end of EGS operation for case 9 (Upper row) and case 10 (Lower row). Left column: 3D plot; Middle: 2D contour plot on a diagonal plane in reservoir; Right: iso-value surface (γ_L=0.5).

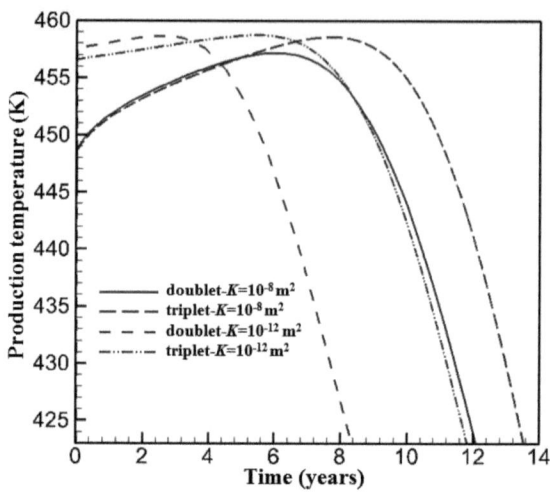

Figure 16 Production temperature curves for EGSs of different well layouts.

4 Well Layout Design

The procedures for constructing an EGS include geological investigation for site-picking, well-drilling, reservoir stimulation, construction of fluid-circulation system, construction of earth-surface power station, and installation of power transmission lines. Well-drilling is a requisite and the most costly procedure, as evident in many geothermal projects [4]. It was reported that the cost of well-drilling could amount to 50%-60% or more of the total capital investment [14, 67]. The well-drilling technology is relatively mature as it has been developed and applied in the oil and gas industry for decades [68, 69]. However, geothermal drilling, especially for applications in EGS, is often far more difficult than in the oil and gas operations as the HDR is usually harder and of higher temperature, and the fluids may be corrosive to the drill bit as well [14, 67-70]. New technologies of borehole drilling are critically needed in the development of commercially viable EGS. A proper design of well layout may reduce technical risks at well-drilling and bring positive effects on the economic performance of EGS.

The simulation results presented in the foregoing subsection 3.6 has already demonstrated very strong well layout effects of EGS heat extraction performance. In this section, we will carry out a more detailed study on EGS well layout effects and attempt to propose some basic principles for EGS well layout design.

4.1 Cases considered

We consider four well layouts, as sketched in Fig. 17, for a thorough study to the relevant well layout effects. Besides the standard doublet well layout, we have designed two triplet well EGSs with one injection well and two production wells, and a quintuplet well EGS with one injection well and four surrounding production wells. Particularly for the triplet well EGSs, in terms of the relative positions of the three wells, there are two well layouts considered. One has a triangular arrangement about the three wells (referred as triplet-triangle hereinafter), as shown in Fig. 17c; the other has the three wells aligned along a straight line (referred as triplet-straightline hereinafter), as shown in Fig. 17b.

The reservoir considered in this section is a 500×500×500 m cubic volume, centered at subsurface 4000 m depth. The assumed geothermal gradient is 4 K/100 m and the ground temperature is 300 K, then it is easy to calculate the average temperature in the reservoir, 460 K. This geothermal resource can meet the general temperature demand of an ordinary geothermal power plant.

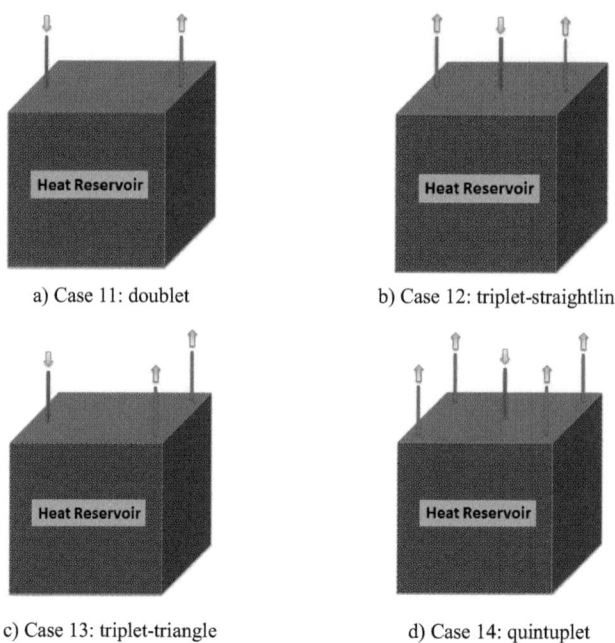

a) Case 11: doublet   b) Case 12: triplet-straightline

c) Case 13: triplet-triangle   d) Case 14: quintuplet

Figure 17 Schematic of the four well layouts considered.

The geometry, including geometrical dimensions of the standard doublet (one injection well and one production well) well EGS is displayed in Fig. 18. The distance from the well center to the nearby reservoir boundary is 50 m. The simulated domain is a 2,000×6,000×2,000 m volume. The injection and production wells are both 0.2×0.2 m square-shaped on the xy-plane. In Fig. 18 also presented is the numerical mesh system. Structural hexahedral meshes are used to discretize the whole domain and the meshing process is elaborately controlled to ensure sufficient fine meshes in the injection and production wells and in the reservoir. Totally, there are about 270,000 numerical elements. Grid-independence tests have been conducted to guarantee the present mesh system gives solutions of satisfying accuracy.

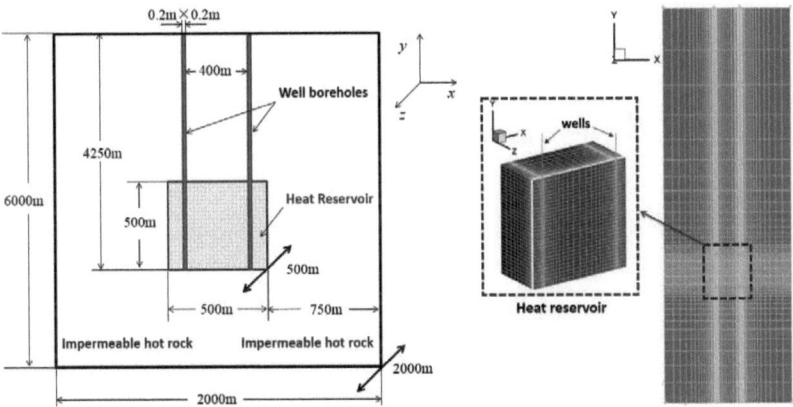

Figure 18 Geometry (including geometrical dimensions) and mesh of the doublet well EGS.

This section still takes liquid water as the heat transmission fluid and assumes no phase change occurring during the subsurface heat extraction process. Thermophysical properties of the fluid and rock are constant, not changing with temperature and/or pressure, which have been already listed in Table 1.

In this section, we assume that the EGS reservoir has been homogeneously fractured, being of constant and uniform porosity and permeability, which are 0.01 and $10^{-14}\,\mathrm{m}^2$, respectively. We consider all the four cases (see Fig. 17) have the same fluid injection rate, 50 kg/s, while the single-well fluid productivity varies from 12.5 to 50 kg/s owing to different number of production well(s), as tabulated in Table 3. Cases differ from each other due to the reservoir well layout and/or the single-well productivity. The parameters, h and a, reflect actually the constitutive condition for heat exchange between the fluid flowing in the fractures and the rock matrix in the reservoir. We consider the product of h and a as a single parameter and specify it to be 1.0 W/m³/K for all the four cases. Initially, the injection and production wells are full of water of temperature 300 K; the temperature of water in the fractures of the reservoir is equal to the local rock matrix temperature. The inflow water temperature is fixed at 343.15 K. For the fluid flow, the inlet boundary condition is fixed mass flow rate and the outflow boundary condition fixed fluid pressure.

Table 3 Cases studied in Section 4

Case #	Well layout	Mass flow rate Q (kg/s)	Single-well production Qs (kg/s)
11	Doublet	50	50
12	Triplet-straightline	50	25
13	Triplet-triangle	50	25
14	Quintuplet	50	12.5

In the following sub-sections 4.2-4.6, we will present simulation results to analyze the detailed well layout effects, to propose the principles of well layout design, and to discuss the recoverable HDR heat. To facilitate analyses and discussion, we define six parameters relevant to the EGS heat extraction process.

1) Production temperature $T_{f,out}(t)$: the fluid temperature at the outlet of the production well.

2) EGS abandonment temperature $T_{f,a}$: the outflow fluid temperature being 10 K lower than the maximum production temperature, different from its general definition that is referred to the average rock temperature in the reservoir being 10 K lower than

44

its initial value [71-73]. For example, if the maximum production temperature is 460 K, the EGS abandonment temperature should be 450 K, allowing 10 K temperature drop of the outflow fluid. We put forth this definition of EGS abandonment temperature mainly due to the reason that the performance of relevant equipments is affected directly by the production temperature [74], instead of the rock temperature in the reservoir. Note that this definition of EGS abandonment temperature $T_{f,a}$ is different from that defined in Section 3.

3) EGS service-time or lifetime τ: the time-duration for an EGS being operated until the production temperature $T_{f,out}(t)$ declines down to the EGS abandonment temperature $T_{f,a}$.

4) Local heat extraction ratio $\gamma_L(t)$: the extracted heat divided by the stored heat, locally. The definition of γ_L can be expressed as

$$\gamma_L(t) = \frac{T_{s,i} - T_s(t)}{T_s(t) - T_o} \tag{9}$$

where $T_{s,i}$, $T_s(t)$ and T_o represent the initial local rock temperature, the local rock temperature at time instant t, and the ground surface temperature, respectively. Note that this definition of local heat extraction ratio is the same as that defined in Section 3.

5) Overall heat extraction ratio $\gamma(t)$: the volumetric average heat extraction ratio in the reservoir, that is,

$$\gamma(t) = \frac{\int_{V_R} \gamma_L(t) dv}{V_R} \tag{10}$$

where V_R represents the reservoir volume.

6) Proportion of thermal compensation from rocks enclosing the reservoir, $\beta(t)$: the heat extracted from the surrounding impermeable hot rocks divided by the accumulative heat extraction amount of the fluid, that is

$$\beta(t) = \frac{\int_{V_s}[T_{s,i} - T_s(t)](\rho c_p)_s dV}{\int_0^t Q[T_{f,out}(t') - T_{f,in}](\rho c_p)_f dt'} \tag{11}$$

where V_s is the volume of the rocks enclosing the reservoir and Q the volumetric flow rate of the heat transmission fluid. V_s is physically infinite, but practically only the rocks that are close enough to the heat reservoir will have detectable contribution to the heating of the heat transmission fluid.

4.2 Heat extraction of doublet EGS

Preferential or short-circuit flow in reservoir is a notorious issue annoying EGS researchers and engineers [7, 31, 57]. From the x-velocity distribution, shown in Fig. 19, we see an obvious preferential flow exists in the reservoir of the doublet EGS, i.e. case 11. The fluid prefers to flow in a narrow-z region centering about the mid-z xy-plane.

Figure 20 shows the temperature of rock in the reservoir at four time instants. Upon the EGS operation, the injected cold fluid quickly cools down the rock mass adjoining to the injection well borehole and a low rock temperature region forms therein. As the heat extraction process progresses, the low-temperature region gradually expands.

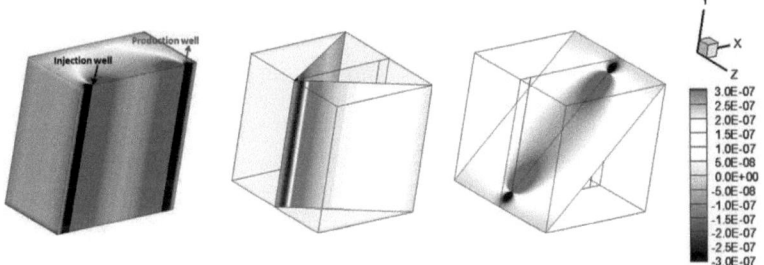

Figure 19 *x*-velocity (m/s) distribution in the heat reservoir of the doublet well EGS (case 11). Left: 3D distribution in half of the reservoir geometry; middle: contour plots on two representative planes; right; contour plot on a diagonal plane.

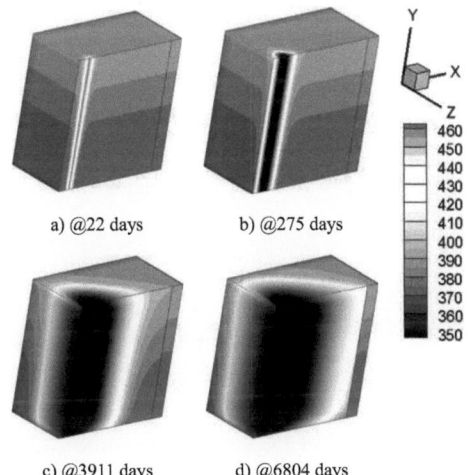

a) @22 days   b) @275 days

c) @3911 days   d) @6804 days

Figure 20 Temperature (K) of rock in the heat reservoir of the doublet well EGS (case 11).

The production temperature as a function of EGS operation time is displayed in Fig. 21. If the fluid can be fully heated-up by the rock, the production temperature is around 460 K, which is the initial average rock temperature in the reservoir. Once the low rock temperature region expands too close to the production well, i.e. at about 8 years into the EGS operation, the fluid does not have sufficient time to extract heat

47

from the rock mass in the reservoir and the production temperature begins to decrease. From Fig. 21 we determine as well the lifetime of this EGS is 19.6 years, i.e. τ=19.6 years.

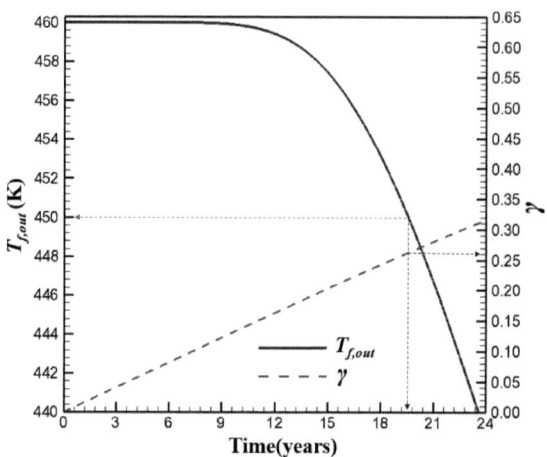

Figure 21　Production temperature and heat extraction ratio curves for the doublet well EGS (case 11).

We calculated the heat extraction ratio, $\eta_t(t)$ and $\chi(t)$. Figure 22 depicts the calculated $\eta_t(\tau)$ results. At the end of EGS operation (time=τ), a large portion of heat stored in the reservoir has not been extracted. From the overall heat extraction ratio curve as a function of EGS operation time, $\chi(t)$, which has been already depicted in Fig. 21, we find the overall heat extraction ratio at the end of EGS operation, i.e. $\chi(\tau)$, is only about 0.26. That is to say, about 74% of the total heat stored in the heat reservoir is not mined out.

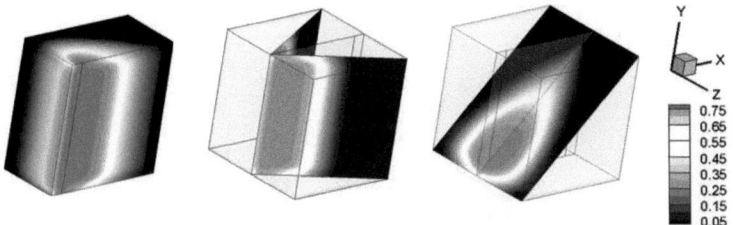

Figure 22   The final local heat extraction ratio distribution, $\gamma_L(\tau)$, in the heat reservoir of the doublet well EGS (case 11). Left: 3D distribution in half of the reservoir geometry; middle: contour plots on two representative planes; right; contour plot on a diagonal plane.

The preferential flow makes most of the fluid has limited residence time in the reservoir and a large portion of rock mass has little chance to access the heat transmission fluid. The occurrence of preferential flow deteriorates the heat extraction performance and leads to a premature EGS operation.

During EGS operation, the rock mass in the reservoir is cooled by the heat transmission fluid, and a temperature difference is thus formed between the rock matrix in the porous reservoir and the rock enclosing the reservoir. Heat is then conducted from the surrounding impermeable rocks to the reservoir. This thermal compensation action is illustrated in Fig. 23. We see strong thermal compensation action in the vicinity region of the injection well, where the rock mass has been sufficiently cooled down by the injected fluid. We quantify this thermal compensation effect on EGS heat extraction process by the parameter $\beta(t)$, defined by Eq. (11). It is calculated that at the time instant τ (i.e. 19.6 years), β amounts to be 6.3%, meaning accumulatively, about 6.3% of the heat extracted by the outflow fluid has come originally from the rock mass enclosing the reservoir.

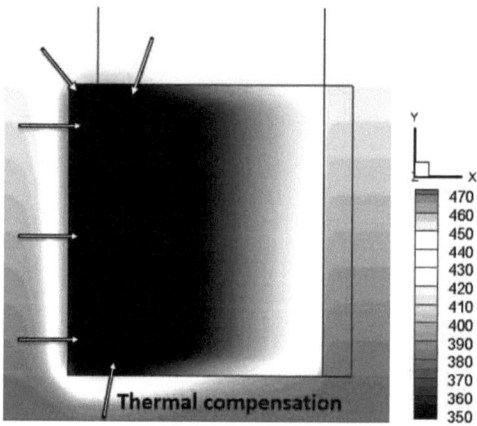

Figure 23 Rock temperature (K) on the central *xy*-plane at the end of the EGS operation for case 11.

4.3 Detailed well layout effects

During design and practical development of EGS, enhancing subsurface inter-well connectivity and alleviating preferential flow in the reservoir are two important and seemingly conflicting measures that deserve full consideration for achieving better EGS performance. Current reservoir stimulation technologies have not yet reached that high level that can create a reservoir of desired fracture networks with ease [19, 75, 76]. The strategy of arranging more than one production well can probably give consideration to both measures aforementioned and has been implemented in a few EGS power stations [6, 7, 77]. How does the well layout affect the EGS heat extraction performance and what are the underlying fundamentals? In sub-section 3.6, we have presented some preliminary results about this. In the following paragraphs of this sub-section, we attempt to answer these questions with more details.

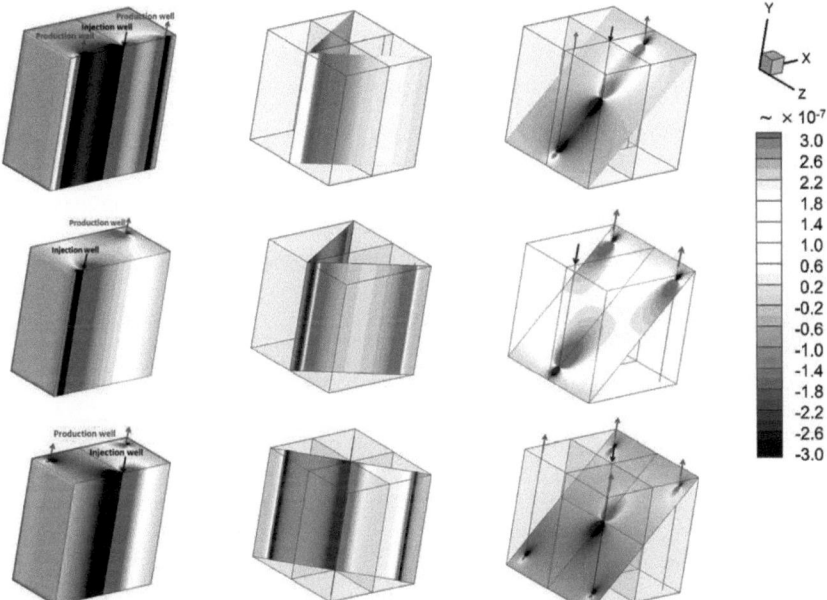

Figure 24 x-velocity (m/s) distribution in the reservoir for cases 12, 13 and 14. Upmost row: case 12; mid-row: case 13; bottom row: case 14. Left column: 3D distribution in half of the reservoir geometry; middle column: contour plots on two representative planes; right column; contour plot on a diagonal plane.

Results of fluid flow field in the reservoir for cases 12, 13, and 14 are shown in Fig. 24. The fluid flow pattern in the reservoir is largely influenced by the well layout. Compared with the corresponding results for case 11 (see Fig. 19), the triplet-straightline well layout (case 12) evidently aggravates the preferential flow and makes the fluid flow being more confined in a narrower-z region centering about the mid-z xy-plane; the triplet-triangle (case 13) and quintuplet (case 14) EGSs both show evident improvement at the fluid flow distribution in the reservoir, in particular for the triplet-triangle well EGS (case 13), the fluid flow distributes quite uniformly in the reservoir.

Figure 25 presents the production temperature as a function of the EGS operation time for all the four cases. The curves differ from each other mainly at the time duration that the production temperature retains at the maximum production temperature, i.e. about 460 K. Case 13 shows the longest time duration, case 14 the second longest, case 11 the third, and case 12 the shortest. As mentioned in Section 4.1 in relation with Fig. 21, the decrease of production temperature is caused by the injected cold fluid breaks through the reservoir and does not have enough time to extract heat from the rock mass. Therefore, the fluid flow pattern in the reservoir dictates the evolution of production temperature. More uniform flow distribution or less preferential flow in the reservoir leads to better EGS performance. It is easy to determine from Fig. 25 that the EGS lifetimes of the four cases, cases 11, 12, 13, and 14, are 19.6, 9.8, 29.8, and 26.9 years, respectively.

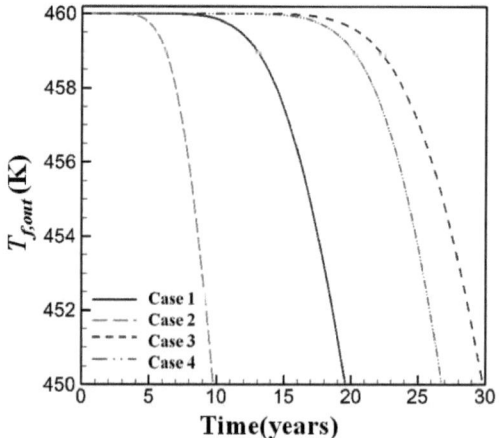

Figure 25   Evolution of EGS production temperature for the four cases: case 11-14.

Results shown in Fig. 25 indicate that an EGS of triplet well layout perform better than an EGS of doublet well layout only if the triplet well layout has been properly designed, and an EGS of quintuplet well layout may even perform worse than an EGS of triplet well layout. We further deduce from the calculated results that simply drilling more wells does not surely enhance the EGS performance as the well layout may play a more determinant role, not even mentioning drilling more wells may significantly increase the initial investment of EGS plants.

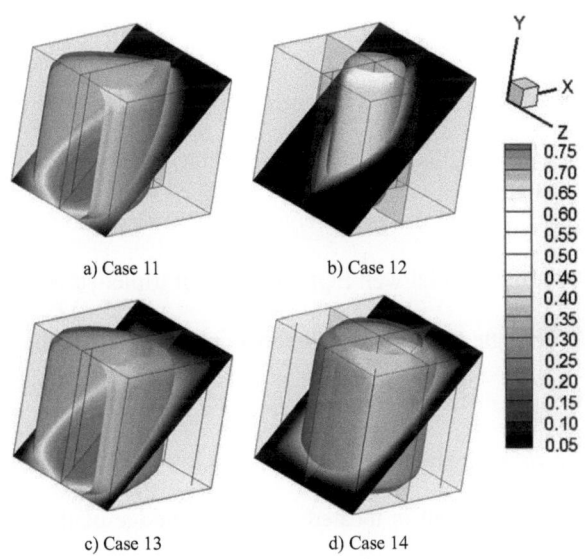

Figure 26   The final local heat extraction ratio distribution in the reservoir and an iso-surface with $\chi(\tau) = 0.4$ for all the four cases.

We calculate the local heat extraction ratio in the reservoir and present its distribution on a representative plane of the reservoir at the end of EGS operation in Fig. 26 for all the four cases. Specially, we draw additionally iso-value surfaces with

$\gamma_i(\tau) = 0.4$ in this figure. These plots clearly show that the heat extraction process of case 13 has been carried out with the best completeness, case 14 the second best, case 11 the third, and case 12 the worst. All the four cases are seen to have some regions with very low local heat extraction ratio. Generally, the low heat extraction ratio regions are close to the production well or wells.

Figure 27 summarizes the calculated overall heat extraction ratio and thermal compensation proportion at the end of EGS operation, i.e. $\gamma(\tau)$ and $\beta(\tau)$ for all the four cases. It is seen that the final overall heat extraction ratio for the EGSs with different well layouts varies within 0.136 to 0.397. The EGS of triplet-straightline well layout has the worst heat extraction performance, the doublet well EGS the second worst, the quintuplet well EGS the third worst, and the EGS of triplet-triangle well layout the best. The final thermal compensation proportions are 0.063, 0.020, 0.075, and 0.035 for the doublet well EGS, the EGS of triplet-straightline well layout, the EGS of triplet-triangle well layout, and the quintuplet well EGS, respectively. The doublet well EGS and the EGS of triplet-triangle well layout get relatively larger amount of thermal compensation from the rocks enclosing the reservoir mainly due to the fact that the injection well is located close to the edge of the reservoir. During EGS operation, a low rock temperature region is formed in the vicinity region of the injection well (Refer to Figs. 20, 22 and 26). Positioning the injection well close to the edge of the reservoir thereby facilitates the thermal compensation process (Refer to Fig. 23). The EGS of triplet-triangle well layout gets slightly more thermal compensation than the doublet well EGS as the former has longer lifetime. The same reason leads to the slight difference (0.015) of thermal compensation proportion between the quintuplet well EGS and the EGS of triplet-straightline well layout.

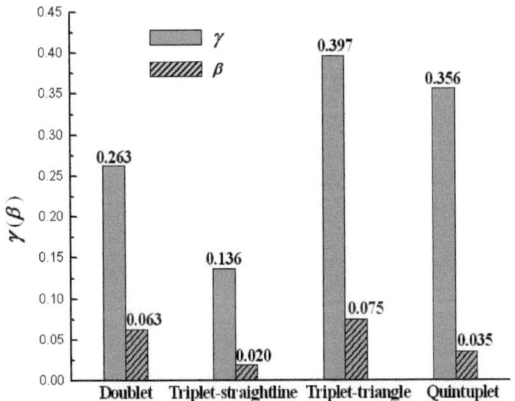

Figure 27   Overall heat extraction ratio and thermal compensation proportion at the end of EGS operation for all the four cases.

4.4 Design of well layout

From the results detailed in Sub-sections 4.2 and 4.3, we see that multiplet well layout does have positive effects on the heat extraction of EGS if the well layout is properly designed. The analyses in the foregoing sub-sections also indicate that at least two basic principles need to be followed during the design of EGS well layout: 1) longer major flow path; 2) less preferential flow. Figure 28 illustrates the length of major flow path in reservoirs of EGSs with different well layouts. It is seen that the triplet-straightline well layout simply reduces half of the major flow path of the doublet well EGS, the other two cases both have longer major flow path than the doublet well layout, and the triplet-triangle well layout has the longest flow path. In addition, to maximize the thermal compensation from rocks enclosing the reservoir, the injection well needs to be positioned close to the edge of the reservoir. As the triplet-triangle well layout strictly follows all the principles, it gives the best heat extraction performance.

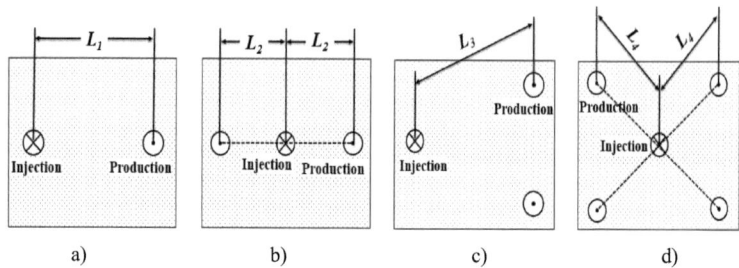

a)   b)   c)   d)

Figure 28   Length of major flow path, a) the doublet well layout, b) triplet -straightline well layout, c) triplet-triangle well layout, and d) quintuplet well layout. $L_2 = 0.5\ L_1$, $L_3 = 1.12\ L_1$, $L_4 = 0.71\ L_1$.

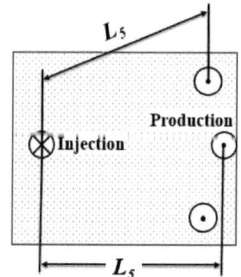

Figure 29   Schematic of the optimized quartuplet well layout, $L_5 = L_3 = 1.12\ L_1$.

Accordingly, we further design a quartuplet well layout (schematic displayed in Fig. 29) and simulate the long-term operation of this quartuplet well EGS. Its lifetime is calculated to be 30.6 years and the final overall heat extraction ratio is 0.408. Both are a little better than those of case 13, validating the basic principles we proposed for the design of EGS well layout. It is worth pointing out that the improvement at the EGS heat extraction performance is not significant compared to the EGS of triplet-triangle well layout, indicating the triplet-triangle well layout is already a very good well layout design.

4.5 Estimation of recoverable HDR resource

The HDR heat resource is ubiquitous across the planet. Nevertheless, there are some places, which may be not suitable for EGS construction due to social and/or humanistic factors, such as national parks, recreation areas, urban areas, major highways, utility corridors, and national monuments. Besides, the regions where the geothermal gradient is relatively low, the areas where the underground rock is too stiff to drill, and/or the places where water (i.e. the common heat transmission fluid) is very scarce are all not appropriate choices for EGS construction from a commercial perspective. To estimate the potential of HDR heat resource, all these factors must be taken into account. More importantly, both the fraction of heat that can be extracted from EGS reservoir and the heat-to-electricity conversion coefficient provide the most compelling information for the estimation. A recovery factor of 2%, which may be somewhat conservative, was used to estimate the potential of HDR heat resource in US [4].

Since the total heat reserve in HDR is constant, the heat extraction ratio becomes the most important factor that influences the estimation of EGS potential [78]. Considerable efforts [71-73] have been expended to estimate the HDR heat recovery factor. Sanyal and Butler [71] used a 3-dimensional numerical model and calculated the fraction of HDR heat that could be extracted. They found that the recoverable heat from a minimum 1×10^8 m^3 (approximately 500×500×500 m dimensions) reservoir volume is within 34% - 47% of the total heat stored, and they asserted that this recovery factor is independent of well layout, fracture spacing, and reservoir permeability, as long as the reservoir volume exceeds 1×10^8 m^3. Grant and Garg [72] expressed their doubts on Sanyal and Butler's results as the used model by Sanyal and Butler encompassed too many simplifications. Williams [73] also suggested a much different HDR heat recovery factor, less than 0.1. To date, there are yet no field test data on the long-term heat extraction process of EGS, which makes validation and calibration of the numerical results extremely difficult. Nevertheless, numerical simulation based on rational simplification models may still be the most effective method for the estimation of HDR recoverable heat.

We take the reservoir as a homogeneous porous medium and the overall heat extraction ratio, defined by Eq. (10), at the end of EGS operation, i.e. $\chi(\tau)$, is actually the HDR heat recovery factor. Besides the 4 cases listed in Table 3, we simulated 18 cases more. The calculated recovery factors, summarized in Table 4, are found to be within a range of 13%-49%, strongly dependent on the well layout of EGS, whereas relatively less dependent on the other parameters examined. If considering a 10% heat-to-electricity conversion efficiency [79] and an additional reduction factor of 0.95 as some places may be not suitable for constructing EGS plants, we estimate the potential of HDR heat for electricity-generation is only about 1.2% -4.9 % of the total heat storage.

Table 4 Cases considered for the estimation of HDR recoverable heat and the obtained results.

Case #	Well layout	K (m²)	∇T (K/100m)	Q (kg/s)	$T_{f,in}$ (K)	τ (year)	γ (%)	$\beta(\tau)$ (%)
11	Doublet	10^{-14}	4	50	343	19.6	26.3	6.3
12	Triplet-straight line	10^{-14}	4	50	343	9.8	13.6	1.96
13	Triplet-triangle	10^{-14}	4	50	343	29.8	39.7	7.50
14	Quintuplet	10^{-14}	4	50	343	26.9	35.6	3.5
15	Doublet	10^{-14}	4	150	343	6.34	26.0	2.06
16	Doublet	10^{-10}	4	150	343	6.7	26.1	2.12
17	Doublet	10^{-12}	4	150	343	6.34	26.0	2.06
18	Doublet	10^{-16}	4	150	343	6.34	26.0	2.06
19	Triplet-straight line	10^{-14}	4	150	343	3.11	12.44	0.87
20	Triplet-triangle	10^{-14}	4	150	343	9.8	38.5	3.78
21	Quintuplet	10^{-14}	4	150	343	8.58	35.2	1.73
22	Triplet-straight line	10^{-14}	4	100	343	4.76	12.83	1.20
23	Triplet-triangle	10^{-14}	4	100	343	14.85	48.9	4.94
24	Quintuplet	10^{-14}	4	200	343	6.38	33.33	1.39
25	Doublet	10^{-14}	4	50	300	18.3	34.65	5.20
26	Triplet-straight line	10^{-14}	4	50	300	8.95	15.79	1.86
27	Triplet-triangle	10^{-14}	4	50	300	28.35	48.55	7.32
28	Quintuplet	10^{-14}	4	50	300	24.24	44.32	3.40
29	Doublet	10^{-14}	5	50	343	18.33	25.9	5.2
30	Triplet-straight line	10^{-14}	5	50	343	8.99	13.2	1.8
31	Triplet-triangle	10^{-14}	5	50	343	28.43	38.6	6.7%
32	Quintuplet	10^{-14}	5	50	343	24.32	34.3	3.2%

5 Conclusions

We detailed a three-dimensional transient model for subsurface thermo-hydraulic process in EGS. This model has a couple of salient features. It treats the porous heat reservoir as an equivalent porous medium while considers local thermal non-equilibrium between solid rock matrix and fluid flowing in the fractures and employs two energy conservation equations to describe heat transfer in the rock matrix and in the fractures, respectively. In this way, the model makes up the major disadvantage (i.e., incapability of modeling local heat exchange between rock matrix and fluid) of conventional thermal model for porous media. The other salient feature of this model is its capability of simulating the complete subsurface thermo-hydraulic process in EGS. The EGS subsurface geometry of interest physically consists of multiple domains while computationally we treat it as a single-domain of multiple sub-regions associated with different sets of characteristic properties (porosity and permeability etc.). This circumvents typical difficulties about matching boundary conditions between sub-domains in traditional multi-domain approaches and facilitates numerical implementation and simulation of the complete subsurface heat exchange process.

With this model, a comprehensive study to EGS heat extraction process was performed. Owing to the application of thermal non-equilibrium model for heat transfer in porous heat reservoir, the obtained results provide insightful information on the local heat exchange process occurring in EGS heat reservoir. There exist apparent temperature differences between the solid rock matrix and heat transmission fluid in the reservoir during EGS heat extraction, justifying the use of thermal non-equilibrium model.

Analyzing the simulated fluid flow fields in homogeneously fractured heat reservoirs of various permeability values unravels the formation mechanism of preferential (or short-circuit) flow. During fluid circulation, a basic pressure difference caused by the gravity effect and the transformation between hydrodynamic and hydrostatic pressure, and the Darcy resistance imposed by the porous reservoir are the two major factors that dictate the flow pattern in the reservoir. For an EGS of larger reservoir permeability, the dominating factor is the basic pressure difference. Cases of different reservoir permeability generally show distinct preferential flow patterns as the Darcy resistance is inversely proportional to the reservoir permeability and the increased Darcy resistance may take the dominating role to dictate the flow pattern. Nevertheless, for an EGS of homogeneously fractured reservoir considered in the present work, the flow pattern changes little once the reservoir permeability is decreased below one Darcy.

The heat extraction performance of EGS is found to be tightly related to the fluid flow pattern in reservoir. Preferential flow is generally detrimental to EGS performance since it leads to non-uniform heat extraction and a premature dropdown at the production temperature. Increasing the reservoir permeability generally improves the flow distribution in the reservoir and can largely prolong the EGS lifetime. An EGS of better heat exchange conditions in reservoir has relatively longer lifetime and the heat exchange process occurring in the reservoir behaves more like the process described by the conventional local thermal equilibrium model.

For an EGS with operation time within ten years, the thermal compensation from rocks surrounding the reservoir is found to contribute little to the heat extraction process. During the early period of EGS operation, this thermal compensation can even have a negative effect on the heat extraction process, which is deemed to be caused originally by the prescribed geothermal gradient. The EGS well layout is found to have profound effects on its heat extraction performance.

Furthermore, we carried out a series of numerical simulations to evaluate and analyze the effects of well layout on EGS heat extraction performance. For the four cases of distinct well layouts considered, the triplet-triangle well EGS shows the best heat extraction performance. A detailed analysis to the simulation results revealed the underlying mechanisms. The triplet-triangle well layout effectively restrains preferential flow in the reservoir and at the same time keeps the major flow path sufficiently long. The thermal compensation from un-fractured rocks surrounding the reservoir contributes some to heat the circulating fluid; about a few percent of the cumulative heat extraction amount can come from the heat stored in these rocks. Moreover, positioning the injection well close to the edge of the reservoir effectively facilitates the thermal compensation process.

Specially, for the EGS considered, which is of a 500×500×500 m homogeneously fractured reservoir, the triplet-triangle well layout may be the best choice for heat extraction, since even with an optimized quartuplet well layout, little improvement at the heat extraction performance can be achieved. The HDR heat recovery factors calculated from 22 cases are found to be within 13%-49%, which show strong dependence on the well layout, whereas relatively slight dependence on the parameters like the reservoir permeability, the geothermal gradient, the fluid flow rate, and the fluid injection temperature. Accordingly, we estimate the potential of HDR heat for electricity-generation is about 4.9 % maximized.

For real heterogeneous heat reservoirs, the obtained heat recovery factors may be exaggerated to some extent and the optimized well layout may be significantly different from the triplet-triangle well layout or quartuplet well layout proposed in the present work. Nevertheless, the proposed principles: longer major flow path, less preferential flow, and positioning the injection well as close as possible to the edge of the reservoir, for the design of advanced well layout should be still effective. It is worth pointing out as well that better heat extraction performance of EGS may be achieved by other measures, for instance, drilling directional or horizontal wells, other than designing multi-well heat extraction mode of particular interest in the present work.

Nomenclature

a specific surface area of aperture network (m^2/m^3)

c_p heat capacity $(J/(kg{\cdot}K))$

g acceleration of gravity (m/s^2)

h convective heat transfer coefficient $(W/(m^2{\cdot}K))$

k thermal conductivity $(W/(m{\cdot}K))$

k_s^{eff} effective thermal conductivity of rock (W/(m•K))

k_f^{eff} effective thermal conductivity of heat transmission fluid (W/(m•K))

K reservoir permeability (m^2)

L distance from injection well to production well (m)

L_1 distance from injection well to production well of case 1 (m)

L_2 distance from injection well to production well of case 2 (m)

L_3 distance from injection well to production well of case 3 (m)

L_4 distance from injection well to production well of case 4 (m)

L_5 distance from injection well to production well of the quartuplet EGS (m)

N_e number of numerical elements (-)

P pressure (Pa)

Q mass flow rate of heat transmission fluid (kg/s)

Q_s single-well production of heat transmission fluid (kg/s)

t time (s)

t' time (s)

T temperature (K)

T_f liquid temperature (K)

$T_{f,a}$ abandonment temperature (K)

$T_{f,in}$ injection temperature (K)

T_o ground surface temperature (K)

$T_{f,out}$ production temperature (K)

T_s rock temperature (K)

$T_{s,i}$ initial rock temperature (K)

u velocity vector (m/s)

v volume (m^3)

V_R reservoir volume (m^3)

V_S volume of rock surrounding the reservoir (m^3)

x horizontal axis in Cartesian coordinates

y vertical axis in Cartesian coordinates

z horizontal axis in Cartesian coordinates

Greek symbols

ρ density (kg/m^3)

ε porosity (-)

μ viscosity (Pa•s)

τ EGS lifetime (years)

β proportion of thermal compensation

γ heat extraction ratio

γ_L local heat extraction ratio

θ time (s)

∇T geothermal gradient (K/100m)

Subscripts/superscripts

eff   effective

i	initial
L	local
f	fluid
s	solid or rock or single well

Acknowledgements

Financial support received from the China National "863" Project (2012AA052802), the CAS "100 talents" Program (FJ), and the China National Science Foundation (51206174) is gratefully acknowledged.

References

[1] World Energy Council. "Survey of Energy Resources." Houston, USA: World Energy Council. (1998).

[2] K.W. Li, Comparison of geothermal with solar and wind power generation systems, Proceedings of the Thirty-eighth Workshop on Geothermal Reservoir Engineering, Stanford University, Stanford, California, 2013.

[3] M. Lacirignola, I. Blanc, Environmental analysis of practical design options for enhanced geothermal systems (EGS) through life-cycle assessment, Renewable Energy 50 (2013) 901-914.

[4] J.W. Tester, B.J. Anderson, A.S. Batchelor, et.al, The future of geothermal energy: impact of enhanced geothermal systems (EGS) on the United States in the 21st Century. Final report to the US Department of Energy Geothermal Technologies

Program, Cambridge, MA: 504 Massachusetts Institute of Technology, 2006.

[5] R. Bertani, Geothermal power generation in the world 2005–2010 update report, Geothermics 41 (2012) 1-29.

[6] A. Genter, K. Evans, N. Cuenot, D. Fritsch, B. Sanjuan, Contribution of the exploration of deep crystalline fractured reservoir of Soultz to the knowledge of enhanced geothermal systems (EGS), Comptes Rendus Geoscience 342 (7) (2010) 502-516.

[7] N. Tenma, T. Yamaguchi, G. Zyvoloski, The Hijiori hot dry rock test site, Japan: Evaluation and optimization of heat extraction from a two-layered reservoir, Geothermics 37 (2008) 19-52.

[8] B.A. Goldstein, A.J. Hill, A. Long, A.R. Budd, F. Holgate, M. Malavazos, Hot rock geothermal energy plays in Australia, Proceedings of the Thirty-fourth Workshop on Geothermal Reservoir Engineering, Stanford University, Stanford, California, 2009.

[9] Z. Wan, Y. Zhao, J. Kang, Forecast and evaluation of hot dry rock geothermal resource in China, Renewable energy 30 (2005) 1831-1846.

[10] U.S. Department of Energy, First Commercial Success for Enhanced Geothermal Systems Spells Exponential Growth for the Future of Geothermal Energy, see: http://www1.eere.energy.gov/geothermal/news_detail.html?news_id=19234.

[11] U.S. Department of Energy. An evaluation of Enhanced Geothermal Systems technology, 2008.

[12] D.E. White, D.L. Williams, Assessment of geothermal resources of the United

States—1975, Geological Survey Circular, Arlington, VA, 1975.

[13] M.W. McClure, Fracture stimulation in enhanced geothermal systems. Master dissertation, Stanford University, Stanford, California, 2009.

[14] Y. Polsky, L. Capuano, J. Finger, Enhanced geothermal systems (EGS) well construction technology evaluation report. SAND2008-7866, Sandia National Laboratories, 2008.

[15] K.K. Bloomfield, P. T. Laney, Estimating well costs for enhanced geothermal system applications, Idaho National Laboratory, Renewable Energy and Power Technologies, Idaho Falls, Idaho 83415, August 2005.

[16] J. Sausse, C. Dezayes, L. Dorbath, A. Genter, J. Place, 3D model of fracture zones at Soultz-sous-Forêts based on geological data, image logs, induced microseismicity and vertical seismic profiles, Comptes Rendus Geoscience 342 (7) (2010) 531-545.

[17] K. Tezuka, H. Niitsuma, Stress estimated using microseismic clusters and its relationship to the fracture system of the Hijiori hot dry rock reservoir, Engineering Geology 56 (1) (2000) 47-62.

[18] G. Zimmermann, G. Blöcher, A. Reinicke, W. Brandt, Rock specific hydraulic fracturing and matrix acidizing to enhance a geothermal system—Concepts and field results, Tectonophysics 503 (1) (2011) 146-154.

[19] M.J. Economides, K.G. Nolte, U. Ahmed, Reservoir stimulation, third ed., Wiley, New York, 2000.

[20] L.N. Germanovich, D.K. Astakhov, M.J. Mayerhofer, J. Shlyapobersky, L.M..

Ring, Hydraulic fracture with multiple segments I. Observations and model formulation, International Journal of Rock Mechanics and Mining Sciences 34 (3) (1997) 97.e1-97.e19.

[21] S. Portier, F.D. Vuataz, P. Nami, B. Sanjuan, A. Gérard, Chemical stimulation techniques for geothermal wells: experiments on the three-well EGS system at Soultz-sous-Forêts, France, Geothermics 38 (4) (2009) 349-359.

[22] M.A. Grant, J. Clearwater, J. Quinão, P.F. Bixley, M.L. Brun. Thermal stimulation of geothermal wells: a review of field data, Proceedings of the Thirty-eighth Workshop on Geothermal Reservoir Engineering, Stanford University, Stanford, California, 2013, pp. 678-684.

[23] J. Taron, D. Elsworth, Coupled mechanical and chemical processes in engineered geothermal reservoirs with dynamic permeability, International Journal of Rock Mechanics and Mining Sciences 47 (8) (2010) 1339-1348.

[24] E.L. Majer, R. Baria, M. Stark, S. Oates, J. Bommer, B. Smith, H. Asanuma. Induced seismicity associated with enhanced geothermal systems, Geothermics, 36 (3) (2007) 185-222.

[25] S.D. Simone, V. Vilarrasa, J. Carrera, A. Alcolea, P. Meier, Thermal coupling may control mechanical stability of geothermal reservoirs during cold water injection, Physics and Chemistry of the Earth 64 (2013) 117-126.

[26] C.E. Bachmann, S.Wiemer, J. Woessner, S. Hainz, Statistical analysis of the induced Basel 2006 earthquake sequence: Introducing a probability - based monitoring approach for enhanced geothermal systems, Geophysical Journal

International 186 (2) (2011) 793-807.

[27] L. Dorbath, N. Cuenot, A. Genter, M. Frogneux, Seismic response of the fractured and faulted granite of Soultz ‑ sous ‑ Forêts (France) to 5 km deep massive water injections, Geophysical Journal International 177 (2) (2009) 653-675.

[28] K.F. Evans, A. Zappone, T. Kraft, N. Deichmann, F. Moia, A survey of the induced seismic responses to fluid injection in geothermal and CO_2 reservoirs in Europe, Geothermics 41 (2012) 30-54.

[29] A.A. Kaniyal, G.J. Nathan, J.J. Pincus, The potential role of data-centres in enabling investment in geothermal energy, Applied Energy 98 (2012) 458-466.

[30] K. Tezuka, H. Niitsuma, Stress estimated using microseismic clusters and its relationship to the fracture system of the Hijiori hot dry rock reservoir, Engineering Geology 56 (1) (2000) 47-62.

[31] R. Parker, The Rosemanowes HDR Project 1983–1991, Geothermics 28(4) (1999) 603-615.

[32] D Brown, R DuTeaux, P Kruger, D Swenson, T. Yamaguchi, Fluid circulation and heat extraction from engineered geothermal reservoirs, Geothermics 28 (4) (1999) 553-572.

[33] B. Sanjuan, J.L. Pinault, P. Rose, A. Gérard, M. Brach, G. Braibant, C. Crouzet, J.C. Foucher, A. Gautier, S. Touzelet, Tracer testing of the geothermal heat exchanger at Soultz-sous-Forêts (France) between 2000 and 2005, Geothermics 35 (5) (2006) 622-653.

[34] A. Polak, D. Elsworth, H. Yasuhara, A. S. Grader, P. M. Halleck, Permeability reduction of a natural fracture under net dissolution by hydrothermal fluids, Geophysical Research Letters 30 (20) (2003) doi:10.1029/2003GL017575.

[35] B. Ledésert, R. Hebert, A. Genter, D. Bartie, N. Clauer, C. Grall, Fractures, hydrothermal alterations and permeability in the Soultz Enhanced Geothermal System, Comptes Rendus Geoscience 342 (7) (2010) 607-615.

[36] B. Ayling, J. Moore, Fluid geochemistry at the Raft River geothermal field, Idaho, USA: New data and hydrogeological implications, Geothermics 47 (2013) 116-126.

[37] G. Zimmermann, I. Moeck, G. Blöcher Cyclic waterfrac stimulation to develop an Enhanced Geothermal System (EGS)—Conceptual design and experimental results, Geothermics 39 (1) (2010) 59-69.

[38] M. Magliocco, T.J. Kneafsey, K. Pruess, S. Glaser, Laboratory experimental study of heat extraction from porous media by means of CO_2, Proceedings of the Thirty-sixth Workshop on Geothermal Reservoir Engineering, Stanford University, Stanford, California, 2011, pp. 704-710.

[39] L.P. Frash, M. Gutierrez, Development of a new temperature controlled true-triaxial apparatus for simulating enhanced geothermal systems (EGS) at the laboratory scale, Proceedings of the Thirty-Seventh Workshop on Geothermal Reservoir Engineering, Stanford University, Stanford, California, 2012, pp. 682-692

[40] A. Bataillé, P. Genthon, M. Rabinowicz, B Fritz, Modeling the coupling between

free and forced convection in a vertical permeable slot: Implications for the heat production of an Enhanced Geothermal System, Geothermics 35 (5) (2006) 654-682.

[41] J. Taron, D. Elsworth, Thermal–hydrologic–mechanical–chemical processes in the evolution of engineered geothermal reservoirs, International Journal of Rock Mechanics and Mining Sciences 46 (5) (2009) 855-864.

[42] T. Xu, E. Sonnenthal, N. Spycher, K. Pruess, TOUGHREACT—A simulation program for non-isothermal multiphase reactive geochemical transport in variably saturated geologic media: Applications to geothermal injectivity and CO_2 geological sequestration, Computers & Geosciences 32 (2) (2006) 145-165.

[43] R. Gelet, B. Loret, N. Khalili, A thermo-hydro-mechanical coupled model in local thermal non-equilibrium for fractured HDR reservoir with double porosity, Journal of Geophysical Research 117 (2012) B07205–B07228 .

[44] A.R. Shaik, S.S. Rahman, N.H. Tran, T. Tran, Numerical simulation of Fluid-Rock coupling heat transfer in naturally fractured geothermal system, Applied Thermal Engineering 31 (10) (2011) 1600-1606.

[45] Y.S. Wu, H.H. Liu, G.S. Bodvarsson, A triple-continuum approach for modeling flow and transport processes in fractured rock, Journal of Contaminant Hydrology 73 (1) (2004) 145-179.

[46] E. Huenges, P Ledru, Geothermal energy systems: exploration, development, and utilization-Chapter 5. Geothermal Reservoir Simulation. Wiley-VCH, 2010.

[47] M. Sahimi, Flow and transport in porous media and fractured rock, second ed.,

Wiley-VCH, 2012, pp. 5-8.

[48] M.J. O'Sullivan, K. Pruess, M.J. Lippmann, State of the art of geothermal reservoir simulation, Geothermics 30 (4) (2001) 395-429.

[49] F.M. Jiang, L. Luo, J.L. Chen. A novel three-dimensional transient model for subsurface heat exchange in enhanced geothermal systems, International Communications in Heat and Mass Transfer 41 (2013) 57-62.

[50] F.M. Jiang, J.L. Chen, W.B. Huang, L. Luo. A three-dimensional transient model for EGS subsurface thermo-hydraulic process, Energy 72 (2014) 300-310.

[51] J.P. Holman. Heat Transfer. McGraw-Hill Inc, New York (1990).

[52] M.O. Häring, U. Schanz, F. Ladner, B.C. Dyer. Characterisation of the Basel 1 enhanced geothermal system, Geothermics 37 (2008) 469-495.

[53] D.W. Brown. A hot dry rock geothermal energy concept utilizing supercritical CO_2 instead of water. Proceedings of the Twenty-Fifth Workshop on Geothermal Reservoir Engineering, Stanford University, Stanford, California, 2000.

[54] J.P. Latham, J. Xiang, M. Belayneh, H.M. Nick, C.F. Tsang, M.J. Blunt. Modelling stress-dependent permeability in fractured rock including effects of propagating and bending fractures, International Journal of Rock Mechanics and Mining Sciences 57 (2012) 100-112.

[55] I. Moeck, G. Kwiatek, G. Zimmermann. Slip tendency analysis, fault reactivation potential and induced seismicity in a deep geothermal reservoir, Journal of Structural Geology 31 (2009) 1174-1182.

[56] A. Ghassemi, X. Zhou. A three-dimensional thermo-poroelastic model for

fracture response to injection/extraction in enhanced geothermal systems, Geothermics 40 (2011) 39-49.

[57] C. Vogt, C. Kosack, G. Marquart. Stochastic inversion of the tracer experiment of the enhanced geothermal system demonstration reservoir in Soultz-sous-Forêts—Revealing pathways and estimating permeability distribution, Geothermics 42 (2012) 1-12.

[58] A. Karrech. Non-equilibrium thermodynamics for fully coupled thermal hydraulic mechanical chemical processes, Journal of the Mechanics and Physics of Solids 61 (2013) 819-837.

[59] G. Radilla, J. Sausse, B. Sanjuan, M. Fourar. Interpreting tracer tests in the enhanced geothermal system (EGS) of Soultz-sous-Forêts using the equivalent stratified medium approach, Geothermics 44 (2012) 43-51.

[60] D. Crandall, H. Siriwardane, G. Bromhal. Modeling fluids and fractures in geothermal systems. Proceedings of the ASME 2011 5th International Conference on Energy Sustainability, Washington DC, USA, 2011, ES2011-54337.

[61] S. Finsterle, Y. Zhang, L. Pan, P. Dobson, K. Oglesby. Microhole arrays for improved heat mining from enhanced geothermal systems, Geothermics 47 (2013) 104-115.

[62] M. Manga, I. Beresnev, E.E. Brodsky, J.E. Elkhoury, D. Elsworth, S.E. Ingebritsen, D.C. Mays, C.Y. Wang. Changes in permeability caused by transient stresses: Field observations, experiments, and mechanisms. Reviews of

Geophysics 50 (2012) 1-24.

[63] L. Rybach, M. Mongillo. Geothermal sustainability-a review with identified research needs, GRC Transactions 30 (2006) 1083-1090.

[64] S.J. Butler, S.K. Sanyal, A.R. Tait. A numerical simulation study of the performance of enhanced geothermal systems. Proceedings of the Twenty-ninth Workshop on Geothermal Reservoir Engineering, Stanford University, Stanford, California, 2004. SGP-TR-175.

[65] K. Pruess, On production behavior of enhanced geothermal systems with CO_2 as working fluid, Energy Conversion and Management 49 (6) (2008) 1446-1454.

[66] D.B. Fox, D. Sutter, K.F. Beckers, M.Z. Lukawski, D.L. Koch, B.J. Anderson, J.W. Tester, Sustainable heat farming: Modeling extraction and recovery in discretely fractured geothermal reservoirs, Geothermics 46 (2013) 42-54.

[67] E. Barbier. Geothermal energy technology and current status: an overview, Renewable and Sustainable Energy Review 6 (2002) 3-65.

[68] F. Wang, T.X. Ren, F. Hungerford, S. Tu, N. Aziz. Advanced directional drilling technology for gas drainage and exploration in Australian coal mines, Procedia Engineering 26 (2011) 25-36.

[69] B.S. Aadnoy, R. Looyeh. Petroleum Rock Mechanics: Drilling Operations and Well Design. 1st ed. Oxford: Gulf Professional Publishing (2011).

[70] W.A. Elders, G.Ó. Friðleifsson, A. Albertsson. Drilling into magma and the implications of the Iceland Deep Drilling Project (IDDP) for high-temperature geothermal systems worldwide, Geothermics 49 (2014) 111-118.

[71] S.K. Sanyal, S.J. Butler. An analysis of power generation prospects from enhanced geothermal systems. Proceedings World Geothermal Congress, Antalya, Turkey, 2005.

[72] M.A. Grant, S.K. Garg. Recovery factor for EGS. Proceedings of the Thirty-Seventh Workshop on Geothermal Reservoir Engineering, Stanford University, Stanford, California, 2012.

[73] C.F. Williams. Thermal energy recovery from enhanced geothermal systems – evaluating the potential from deep, high-temperature resources. Proceedings of the Thirty-Fifth Workshop on Geothermal Reservoir Engineering, Stanford University, Stanford, California, 2010.

[74] S. Petty, G. Porro. Updated US geothermal supply characterization. Proceedings of the Thirty-Second Workshop on Geothermal Reservoir Engineering, Stanford University, Stanford, California, 2007.

[75] K.F. Evans, A. Genter, J. Sausse. Permeability creation and damage due to massive fluid injections into granite at 3.5 km at Soultz: 1. Borehole observations, Journal of Geophysical Research 110 (2005) B04203.

[76] G. Zimmermann, A. Reinicke. Hydraulic stimulation of a deep sandstone reservoir to develop an Enhanced Geothermal System: Laboratory and field experiments, Geothermics 39 (2010) 70-77.

[77] D.W. Brown, D.V. Duchane. Scientific progress on the Fenton Hill HDR project since 1983, Geothermics 28 (1999) 591-601.

[78] C.R. Chamorro, J.L. García-Cuesta, M.E. Mondéjar, M.M. Linares. An

estimation of the enhanced geothermal systems potential for the Iberian Peninsula, Renewable Energy 66 (2014)1-14.

[79] M. Yari. Exergetic analysis of various types of geothermal power plants, Renewable Energy 35 (2010)112-121.